Queering Tourism

Gay pride parades are often fantastic spectacles of corporeality involving drag queens, buff boys, dykes on bikes, leather bears and other transgressive bodies. Increasingly these parades are forming the basis of a growing international queer tourism market. From Sydney to Rome *Queering Tourism* analyses the paradoxes of gay pride parades as tourist events. The book explores how the public display of queer bodies – how they look, what they do, who watches, and under what regulations – is profoundly important in constructing sexualised subjectivities of bodies and cities. Gay pride parades are annual arenas of queer public culture where embodied notions of subjectivity are sold, enacted, transgressed and debated.

Drawing on extensive collections of interviews, visuals and written media accounts, photographs, advertisements and her own participation in these parades, Lynda Johnston gives a vibrant account of 'queer tourism' in New Zealand, Australia, Scotland and Italy. In each place, *Queering Tourism* looks at how the relationship between the viewer and the viewed produces paradoxical concepts of bodily difference, and considers how queered spaces of gay pride parades may prompt new understandings of power and tourism knowledges.

Lynda Johnston is a senior lecturer in the Department of Geography, Tourism and Environmental Planning at the University of Waikato, Aotearoa/New Zealand. Her research focuses on social/cultural and feminist geography, critical social theory and tourism.

Routledge studies in human geography

This series provides a forum for innovative, vibrant, and critical debate within Human Geography. Titles will reflect the wealth of research which is taking place in this diverse and ever-expanding field.

Contributions will be drawn from the main sub-disciplines and from innovative areas of work which have no particular sub-disciplinary allegiances.

1 **A Geography of Islands**
   Small island insularity
   Stephen A. Royle

2 **Citizenships, Contingency and the Countryside**
   Rights, culture, land and the environment
   Gavin Parker

3 **The Differentiated Countryside**
   Jonathan Murdoch, Philip Lowe, Neil Ward and Terry Marsden

4 **The Human Geography of East Central Europe**
   David Turnock

5 **Imagined Regional Communities**
   Integration and sovereignty in the global south
   James D. Sidaway

6 **Mapping Modernities**
   Geographies of Central and Eastern Europe 1920–2000
   Alan Dingsdale

7 **Rural Poverty**
   Marginalisation and exclusion in Britain and the United States
   Paul Milbourne

8 **Poverty and the Third Way**
   Colin C. Williams and Jan Windebank

9 **Ageing and Place**
   Edited by Gavin J. Andrews and David R. Phillips

10 **Geographies of Commodity Chains**
   Edited by Alex Hughes and Suzanne Reimer

11 **Queering Tourism**
   Paradoxical performances at gay pride parades
   Lynda T. Johnston

Queering Tourism
Paradoxical performances at gay pride parades

Lynda Johnston

First published 2005
by Routledge
2 Park Square, Milton Park, Abingdon, Oxon, OX14 4RN

Simultaneously published in the USA and Canada
by Routledge
270 Madison Ave, New York NY 10016

Routledge is an imprint of the Taylor & Francis Group

Transferred to Digital Printing 2008

© 2005 Lynda Johnston

Typeset in Garamond by Wearset Ltd, Boldon, Tyne and Wear

All rights reserved. No part of this book may be reprinted or
reproduced or utilised in any form or by any electronic, mechanical,
or other means, now known or hereafter invented, including
photocopying and recording, or in any information storage or
retrieval system, without permission in writing from the publishers.

British Library Cataloguing in Publication Data
A catalogue record for this book is available from the British Library

Library of Congress Cataloging in Publication Data
A catalog record for this book has been requested

ISBN10: 0-415-29800-8 (hbk)
ISBN10: 0-415-48210-0 (pbk)

ISBN13: 978-0-415-29800-1 (hbk)
ISBN13: 978-0-415-48210-3 (pbk)

I dedicate this book to my mother, June Edith Johnston 1939–2005, who was always inspirational, strong and proud.

Contents

List of figures   x
Acknowledgements   xi

1 Proud beginnings   1

2 Queer(y)ing tourism knowledges   9

3 Bodies: camped up performances   31

4 Street scenes: tourism with(out) borders   54

5 Sex in the suburbs or the CBD?   77

6 Cities as sites of queer consumption   100

7 Paradoxical endings   122

References   128
Index   140

Figures

1.1	Snow White, Dunedin Festival Parade, 1972	5
3.1	Edinburgh Drumming Girls	35
3.2	Pride Scotland on Princes Street, Edinburgh	38
3.3	The HERO Marching Boys: 'Marcho Men'	43
4.1	Roadside barriers at the Sydney Gay and Lesbian Mardi Gras Parade	59
5.1	Queen Street and Ponsonby Road HERO parade sites	79
5.2	Collage of newspaper articles	91
5.3	HERO Parade's safe sex billboard	95
6.1	Sydney Mardi Gras Map	102
6.2	Wonder Downunder Tours	109
6.3	'God is Gay', World Pride Roma 2000	120

Acknowledgements

This book is based on research that began as a PhD over a decade ago. Many people have helped me to write this book and to all I am deeply grateful. I especially want to thank Robyn Longhurst for her critical insights, support, friendship and enthusiasm. She has been crucial to the development of many ideas in this book. Special thanks go to Kathryn Besio for her astute commentary and who, like Robyn, read through the first draft of this book. Thank you to Richard Bedford and Robin Peace who both put a great deal of energy into earlier versions of this research. The Geography Department at the University of Waikato first made it seem possible to me that I could write about 'queer things' and question knowledge production. Special thanks go to Elaine Bliss, John Campbell, Lex Chalmers, Elsie Ho, Betty-Ann Kamp, Colin McLeay, Jacquie Lidgard, Max Oulton, Diana Porteous and Evelyn Stokes.

While working at the University of Edinburgh, the Feminist Geography Reading Group, which frequently debates issues of power and subjectivities, helped me to extend my critique. I am grateful for the personal and professional support of Hannah Avis, Liz Bondi and Shonagh McEwan. Thank you to the Edinburgh drumming group for letting me play and for taking part in this project! Over the years, in different parade sites, numerous people have taken time to speak with me about their parade experiences, for which I am indebted.

Some of the arguments made here are inspired by Soile Veijola, whose good advice always comes with great humour. There are many scholars whose work has been influential to this book; in particular I would like to thank Jon Binnie, Louise Johnson and Gill Valentine.

Friends and colleagues who have helped and supported me are: Jane Barron, Melanie Gregg, Susan Higgs, Helen Jamieson, Lindsay Knight, Carolyn Michelle, Karen Nairn, Nan Seuffert, Cathy Stephenson, Sarah Turner, Inge Wilhelm, Karen Wilson, Kay Weaver and Yvonne Underhill-Sem.

I must also thank the staff at Routledge: Terry Clague, commissioning editor, and Katherine Carpenter, editorial assistant, for their gentle but prompting emails regarding my progress and for making the production process operate smoothly. Eleanor Holme – accredited indexer and editor – was marvellous and meticulous. Thank you.

An earlier version of Chapter 2 was published in the *Annals of Tourism Research: A Social Science Journal* (2001) as '(Other) Bodies and Tourism Studies'. I would like to thank the editors and publisher for permission to reprint parts of it. I also wish to thank the following for permission to use copyright material. Portions of Chapter 5 first appeared in the *New Zealand Geographer*, 1997, 53, (2), 29–33; parts of Chapter 4 were previously published by Rowman and Littlefield in *Subjectivities, Knowledges and Feminist Geographies: The Ethics and Spaces of Social Science Research* by Bondi et al. (2002). Thanks to Tourism Australia for permission to use their image 'Wonder Downunder' Brochure by Above and Beyond Tours; and the *New Zealand Herald* for permission to use 'Marcho Men' and 'HERO Parade Safe Sex Billboard' images.

Every effort has been made to trace copyright holders, but if any have inadvertently been overlooked they should contact the publishers or me so that the necessary arrangements can be made.

Finally, I'm lucky to have had an extraordinary family – June, Murray, Karen, Richard and Thomas – who have sustained me with love and encouragement. Furry family members need special mention too – Tyla, Vegas, Raglan and Chico – whose daily routines of walks, barks, play, naughtiness and purrs keep a girl happy. Finally, heartfelt thanks go to Tara Kanji for her positive outlook, love and generosity.

Lynda Johnston

1 Proud beginnings

Every year in most western cities gay pride parades transform streets into queer sites of celebration and protest. Visitors to pride parades line the streets to cheer drag queens, dykes on bikes, leather bears, buff boys, marching girls, gay parents with their kids, gay and lesbian school children and many more. This spectacular display has become an important date for the rapidly exanding queer tourism industry, attracting both queer and non-queer tourists. The entertainment, glitz and glamour is only part of the narrative. In the heat of the Parade the history and materiality of homosexual oppression and resistance rubs against regulated spaces subject to planning controls and state intervention. As such, parades are often subject to public debate about which space they may occupy and what kind of displays are deemed appropriate. At the same time, the commercialisation and commodification of queer – through gay pride parades – has fostered the emergence of new forms of neoliberal cultural citizenship. The success of parades such as the Sydney Gay and Lesbian Mardi Gras and the incorporation of the queer celebration into city promotions and mainstream media may be understood as a kind of 'homonormativity', where once transgressive political displays are now corporatised, regulated and controlled. 'Being proud', out and visible, can be politically transgressive as well as about being accepted in neoliberal forms of sexual citizenship.

The purpose of this book is to address the possibilities and paradoxes of gay pride parades. It does so by knitting together three themes. The first is tourism studies' knowledges. The second is gendered and sexualised bodies at gay pride parades. The third is queerness – the ironic, contradictory and transgressive narratives of queer bodies in tourism spaces.

The first theme is epistemological. I am concerned with knowledges that are produced under the title 'tourism studies'. Tourism studies' knowledges are interdisciplinary and travel across many academic conventions. Despite this fluidity of knowledge, or perhaps because of it, there persist attempts to secure knowledge and truth claims based on the self-knowing Cartesian subject. While postmodern claims have destabilised meanings and thrown into doubt ontological certainties, the Cartesian subject persists and remains under theorised in the field of tourism studies.

The lack of critical attention paid to embodiment in tourism studies may stem, in part, from its epistemological foundations. Studies of tourism have loose disciplinary boundaries within the academy and multiple perspectives arise from disparate and wide ranging discourses. This book explores the costs and benefits of this academic placelessness. Moreover, the study of tourism is understood to be a 'risky' business for academics because field work in tourism is often associated with fun, leisure and holidays and hence not perceived as 'real work' that might arise from the hardships of 'serious' fieldwork. Gaining legitimacy in the academy, for tourism studies, is based on the privileging of rational, objective, masculine and heterosexual perspectives. In the urge for a 'legitimate' academic place, tourism studies has tried to exclude, devalue and demean those subjects which have long been associated with embodiment, for example, women and homosexuals.

This book is an attempt to *queer* tourism studies in order to create spaces for precisely these forms of difference and Otherness. It is timely to examine and question some of tourism studies' meta-theoretical conceptions as the field of tourism studies has yet to provide a sustained reflection on the processes of knowledge production. One way to do this is to examine the ways in which certain knowledges disavow and invoke particular kinds of bodies. Feminist theorists argue that knowledge and knowledge production are not only embodied but always gendered and sexualised (Grosz 1993). Central to this book is that all knowledge and all bodies are gendered and sexualised. There are exciting possibilities for the explicit inclusion of gendered and sexualised bodies within the field of tourism studies, as well as an understanding of the processes of production that lead to heteronormative knowledges. This brings me to my second theme around which this book revolves – queer bodies and gay pride parades.

The field of tourism studies has been slow to examine gay and lesbian tourism. The literature that exists tends to focus on advertising trends and industry motivation for new markets. Work on the materiality of gendered and sexualised bodies is strangely absent given that over the past two decades other social sciences have seen a 'body craze' (in geography, for example, see Robyn Longhurst's (2001a) *Bodies: Exploring Fluid Boundaries*). Some social scientists, however, are contributing embodied work to the field of tourism studies either by positioning themselves in their research or paying closer attention to the bodies of their research subjects. In this book I argue that the bodies articulated in tourism studies have tended to be theoretical, discursive and devoid of gendered and sexualised materialities. My discussion of bodies in and around pride parades is, in part, an attempt to mess with these supposedly rational tourism studies' knowledges. My case studies of gay pride parades require an understanding of the categories that define embodied identities. The understanding that bodies are both socially constructed and 'real' matter is of particular importance in my analysis of the paraders and tourists. Bodies in gay pride parades are controversially and blatantly positioned at the intersection of essentialist and social constructionist readings of gendered and sexualised bodies, and hence the interaction between, and engagement of, the natural with the cultural is under consideration in this book. This is one of the reasons why I have turned to queer theorising.

The third theme of this book, then, queerness, infiltrates bodies, theories and the tourist spaces of gay pride parades. Queer theory might be understood to be a poststructuralist, postmodern, and feminist 'tool box' for understanding sexuality. Emerging as an intellectual movement in the 1990s, queer theory centres on the significance of sexuality and gender and their interrelatedness. Queer theory provides ways to understand sexualities such as lesbian, gay or pan-sexual, as well as that which might be 'odd', 'strange', indeed 'queer' (de Lauretis 1991; Jagose 1996). It also constantly destablises taken-for-granted ideas by rejecting any fixed or stable notions of sexuality and gender. Queer theory offers a critical approach to knowledge construction, including gay and lesbian studies, identity politics and even itself (Jagose 1996). Put another way, queer theory demands that social scientists recognise how heteronormativity can blatantly and/or subtly taint the way we think and write about sexuality. Queer theory provides the 'yarn' that entwines and creates different patterns of tourism, bodies and spaces.

Why bodies and tourism?

I became interested in pursuing research on bodies, gender, sexuality and knowledges when I arrived at the Geography Department, University of Waikato, in 1993. I began my first research project on lesbians and domestic spaces. This research considered the constitutive relationship between sexualised bodies and homes (both lesbian homes and parental/heterosexual homes, see Johnston and Valentine 1995). Inspired by a host of feminist geographers, poststructuralists and other writers concerned with bodies, in 1994 I conducted a masters thesis on female body building. A focus on body building helped me understand the complexities of gendered subjectivities. At the completion of this project I began to think about gender *and* sexuality as crucial to the ways in which bodies inhabit space and vice versa. I also started to question why there was a lack of literature on gender and tourism, when, for example, feminist geographers had made a significant impact with studies of gender and space. At about the same time, the seminal article by feminist sociologists' Soile Veijola and Eeva Jokinen (1994) 'The Body in Tourism' appeared in *Theory, Culture and Society*. Their innovative and daring dialogues were for me inspirational. Conducting research into gay pride parades, and thinking about how these are constructed and enacted as tourist spaces, seemed like an exciting opportunity to bring together my interests of gendered and sexualised bodies plus challenge exclusionary knowledge practices.

I have now been going to, participating in, watching and researching pride parades for over a decade. I have attended parades in Aotearoa/ New Zealand, Australia, the United Kingdom, and Italy. Each of these

4 *Proud beginnings*

parades is specific to their particular cultural and spatial contexts. They take their starting points from the New York Stonewall riots which began on the night of 27 June 1969 when police raided a gay bar, called the Stonewall Inn, in Greenwich Village, New York City, USA. Three days and two nights of rioting became emblematic of defiance of compulsory heterosexuality and discursively established the beginning of the gay liberation movement. These riots also represent the first time in US history when gays and lesbians openly protested and resisted police harassment. The assertion and creation of a new sense of identity was produced, one based on pride of being gay (see Barry Adam's (1995) *The Rise of the Gay and Lesbian Movement*, for more information regarding Stonewall). New discourses of pride, defiance and visibility came to eventually dominate public gay and lesbian political landscapes. I have visited Greenwich Village and the Stonewall Inn in New York several times. A gay and lesbian monument consisting of four figures – two male and two female – has been erected in a small Greenwich Village park. This monument stands as a positive symbol of queer lives as well as marking the site of violence against homosexuality.

It is precisely the connection between identity and pride that is both integral and problematic in queer theory and this identity–pride connection highlights one of many paradoxes of this book. Queer exemplifies a mediated approach to categories of identification. Poststructuralist theorising of identity as provisional and contingent has prompted new forms of subjectivities and political organisation. For example, is it possible to march united and proud under one banner when sexualities are understood to be diverse and different? Furthermore, the promotion of parades as tourist events over the past decade or more, has meant that 'being proud', out and visible, is also about entertainment, consumption, neoliberal forms of citizenry, and the establishment of relationships with multiple audiences.

The first New Zealand parade that I participated in was the Dunedin Festival Parade in 1972. I was seven years old and dressed as Snow White. In many ways, the Dunedin Festival Parade stands in stark contrast to gay pride parades. The uncritical adoption of fairy tale performances reinforces heteronormative notions of western white citizenship (see Figure 1.1).

Snow White's life of domestic chores and her wait for the prince was never my idea of a good time. I did, however, like the outfit! Costumes can embody celebration, subjectivity, subversion, fantasy and politics. In some ways, then, this heteronormative parade may be similar to gay pride parades.

I was excited at the thought of experiencing my first gay pride parade: the Auckland Coming Out Day Parade on 3 December 1995. I hired a camcorder and recorded the parade and the people who arrived to watch the parade. The parade started with a welcome and an acknowledgement that it was for everyone. Anyone walking the street could join in. Those who were 'just there' became part of the parade. Accompanying each parade entry was music, usually a steady club beat. I walked with the parade, sometimes on the road in the parade itself, and sometimes on the footpath. In retrospect

Figure 1.1 Snow White, Dunedin Festival Parade, 1972 (source: Author's own collection).

this parade was very different from other parades I attended, in New Zealand and beyond. The boundaries of parade participants and watchers dissolved in the Coming Out Day Parade, whereas in subsequent parades these boundaries were reinforced and actively policed. The Coming Out Day Parade seemed to be directed much more to the 'gay community', had minimal corporate sponsorship, and seemed very 'everyday', banal, almost 'normal'. It was held during the day and had a relaxed feel about it. There were no television coverage, no anti-gay protesters, and it was held in a suburb well known to be queer friendly.

The following parades that I attended seemed much more about spectacle, entertainment and camp sensibilities. As such, these parades had evolved into major carnivalesque celebrations of complex and multi-layered expressions of queerness, consumption and excess. Parades 'down-under' in New Zealand and Australia are held in summer months – February and March – during which warm weather allows for the exposure of more bare flesh (and adds to the sexualisation of parades). The timing of these parades also allows for synergies with summer tourism to Auckland and Sydney. Auckland's HERO Festival – the month long festival of gay events, the parade and the party – has been strategically scheduled to occur just two weeks prior to the Sydney Gay and Lesbian Mardi Gras Parade. Both events attract international and domestic tourists. These parades are held at night and involve elaborate motorised floats, costumes and dance performances. The Sydney Gay and Lesbian Mardi Gras Parade attracts around 800,000 spectators and the parade takes approximately fours hours to pass through well-defined city streets. On a smaller scale, Auckland's HERO Parade attracts around "200,000 onlookers [who pack] both sides of the street" (Moore 1996: 3).

An opportunity to live and work in Edinburgh from 1998 meant that I was able to continue my interest in Pride events. In 1999 I attended Pride Scotland in Edinburgh which marked the thirtieth anniversary of the Stonewall riots. In 2000 I participated in World Pride Roma and in 2001 I attended and performed in Pride Scotland, Edinburgh, in an all-women drumming group that led the parade.

These parades are multi-layered affairs. They all inevitably include public displays by participants in colourful and outrageous clothing. Sexualised performances can be enacted in numerous ways, for example, by people in drag, leather outfits, tight shorts, topless women on bikes, the wearing of sex toys and the parody of heterosexuality. These bodily displays may invoke stereotypical responses by tourists and academics but my hope is that what I have written will prompt others to engage with ideas of bodies, tourism and knowledges.

The methods used to collect data on the 1996 HERO Parade were focus groups and individual in-depth interviews with parade participants and organisers. Also, participant observation at the HERO parade workshop was carried out during the same time that I worked as a parade volunteer for six weeks prior to the parade. A questionnaire was distributed to tourists on the night of the parade, 17 February 1996. During the 1996 Sydney Mardi Gras Parade questionnaires were also distributed. At the 1997 and 1998 HERO Parades, participant observation was carried out and newspaper items have been collected since the first HERO Parade in 1994. At Pride Scotland focus group interviews and individual interviews were conducted and participant observation was carried out at both Pride Scotland parades and World Pride Roma. At all events textual material such as advertisements, posters, festival programmes, video productions and so on were collected. Empirical data were analysed in three ways: data reduction, data displays and conclusion drawing and verification (Miles and Huberman 1994).

Order of events

Chapter 2 begins with an assertion that tourist activities are directly mediated through bodies. I raise epistemological and ontological questions regarding the location of bodies in tourism knowledges. The question 'what are queer bodies?' cannot be answered without thinking about the spaces that make queer bodies. In this way, it is possible to examine bodies as both 'real' and socially constructed through encounters with different tourism spaces. I trace the Othering of the gendered and sexualised body in tourism studies and argue that the body has not simply been absent from these knowledges. Instead, it has been invoked as reason's underside: the irrational, messy, and feminised subject that tourism academics have shied away from. Cartesian dualisms have structured the production of tourism knowledge and these persist. There are signs of an emerging field of tourism research that is beginning to incorporate questions of embodiment into tourist knowledges. I ask if this research alters some of tourism studies' meta-foundations. Does focusing on embodiment prompt new understandings of power, knowledge and tourism spaces? I offer introductions to three aspects of queer theorising that run throughout the empirical chapters, namely, ideas about performativity, camp and abjection. This chapter – indeed the book as a whole – is offered as a starting point for discussions on bodies and tourism epistemologies and ontologies. In the chapters that follow I draw on empirical material from various gay pride parades.

Chapter 3 begins with the body and I focus on one particular type of parade performance, marching groups. Marching groups highlight the mobility of gendered and sexualised embodiment and upset heteronormative understandings of what it means to be male, female and queer. I draw on my experience as a researcher and participant in Pride Scotland 2001 to unravel the complexities of gender, sexuality pride and tourist spaces. Moving from the northern to the southern hemisphere, I then continue my analysis on embodiment but focus on men's marching groups in Sydney Mardi Gras and HERO parades. In both of these case studies I use performativity and camp sensibilities to give weight to the argument that tourist spaces and bodies are mobile and dynamic.

In Chapter 4 I move from bodies to streets in order to consider more thoroughly the relationship between paraders' and tourists' bodies. I examine the instability of discursive and material borders and argue that tourists may have abject reactions to some paraders' bodies. Some bodies on parade, for example, bodies involved in bondage and sadomasochism (s/m) performances, bodies living with HIV/AIDS, and bodies which contest the gendered/sexed corporeal borders, can be partially explained using Iris Young's (1990) notion of cultural imperialism. These bodies are constructed as 'freaks', 'ugly', 'dirty' and are further Othered by some tourists. Julia Kristeva's (1982) notion of 'abjection' is used to understand tourists' and paraders' reactions. There is a fear that some bodies in the parade are too risqué and therefore cannot be trusted in public streets. Contradictions and ironies structure the analysis of the empirical material in this chapter. Spectators are both drawn to, and at the same time, repulsed by bodies on parade. I draw on tourists' responses to a questionnaire in order to argue that the dominant, unmarked position of the heteronormative tourist is maintained through discourses of liberalism. There is potential, however, to disrupt this unmarked position and question the binary between self/Other, tourist/host, and straight/gay by examining specific parade entries, such as Gaily Normal, Rainbow Youth and the Gay Auckland Business Association

(GABA). These groups perform and parody some of the norms of heterosexuality. Hence they have the potential to destabilise dominant meanings of homosexuality and heterosexuality, host and tourist.

A debate over the site of gay pride parades which raged in Auckland, New Zealand's largest city, is the focus of Chapter 5. I draw on data to challenge the notion that sexuality is connected to private spaces. In doing so, I collapse the distinctions usually held between 'real' material places and their representations. Newspaper articles, gay maps and gay magazine publications are woven together with interview and participant observation data. I argue that parades are socially constructed tourist sites which 'queer' the streets. Not only do the parades 'queer' the streets, however; they also foreground the problematic position of 'private' bodies in public places. City councillors, the Mayor and the Deputy Mayor of Auckland were all vocal in their opposition to the parade being held in the central business district of Auckland, despite the popularity of the event for Auckland tourism. The shift of the parade to Ponsonby, the 'gay suburb' of Auckland, highlights the opposition between queer bodies and public spaces. The debate over the parade site illustrates clearly the ways in which western hierarchical dualisms are inscribed on bodies and places.

Chapter 6 extends the book's borders to that of city tourism and the promotion of pride as an event to attract cosmopolitan tourists. It is based on the following cities – Auckland, Sydney, Edinburgh and Rome – and provides a basis from which to examine the paradoxes between claims to equality and entry into sexual citizenry. Throughout the chapter I argue that the performance of city tourist sites reflects and reinforces various hegemonic power relations. Neoliberal discourses of what it means to be a regulated queer citizen frame this chapter.

This book concludes by revisiting conceptualisations of bodies and tourism spaces running through the text. By exploring the ways in which theorisations of sexualised and gendered subjectivities invoke and produce different tourist spaces, I rework the connections between the preceding chapters. The conclusion offers an account of some of the boundaries that are made and broken at gay pride parades. Among the boundaries that are problematised in this book are those around academia and academic knowledge. Research is profoundly influenced by our experiences of negotiating such boundaries, through the choice of this topic and my research methods. I am acutely aware of the dangers of colonising the experiences and knowledges of those on whom my research focuses. But I am also committed to increasing the permeability of the boundaries surrounding academic knowledge, to queering and unsettling the prestige it assumes. The chapters that result therefore engage with knowledges produced in a wide array of settings – bodies, streets, suburbs, cities – through media, street performances, and numerous conversations entered into in the course of this research. By working with these various knowledges I seek to queer ways of thinking about tourism spaces and subjectivity.

2 Queer(y)ing tourism knowledges

> Sydney is the largest city in Australia and arguably the queer capital of the Southern Hemisphere. Sydney's lavish and stylish annual Mardi Gras celebration becomes the focus of the gay world in February/March of each year attracting the best part of a million participants onto the streets.
> (http://gaytravelnet.com/aus/Sydney.html)

Quotations like this one highlight the recent emergence of Sydney as the 'queer capital' down under. Numerous gay travel companies promote tours to Australia to coincide with Mardi Gras. Sydney has become an important destination for gay and lesbian travellers, as well as non-queer tourists. The following newspaper report from the *Sydney Morning Herald* compares and contrasts two separate tourist spectacles: the Chinese New Year celebrations in Hong Kong and the Sydney Gay and Lesbian Mardi Gras Parade. For a married couple – Susan and Peter – the Sydney Gay and Lesbian Mardi Gras Parade appear to be an exceptional tourist event:

> As Susan (middle-aged British visitor) said to her husband Peter (ditto) at the start of Saturday night's cavalcade of camp, 'it's good to have a bit of colour in life, isn't it?' . . . They were in Hong Kong for Chinese New Year in January to see 30-storey skyscrappers lit in electric hues. But it was nothing like this. Nothing like the Sydney Gay and Lesbian Mardi Gras.
> (Hill 1998: 1)

Such is the response of many tourists to the Sydney Gay and Lesbian Mardi Gras Parade. Susan and her husband Peter cheered, roared and shrieked on whistles as the 1998 Sydney Gay and Lesbian Mardi Gras parade passed through the streets (Hill 1998). The parade might be a collective consumer experience, but there exist diverse and competing experiences which revolve around gendered and sexualised subjectivities. The spectacle of bodies on parade for tourists to 'gaze' at is an obvious starting point to discuss the tensions and possibilities of bodies and tourism studies.

Gay travel marketing placed beside Susan and Peter's response to Mardi Gras illustrate complex tourist constructions of pride parades, sexualised bodies and places. Not only are there competing demands and desires of queer groups for pride parades (bringing together often disparate groups under one banner), but there are also tensions and contradictions when pride parades are consumed by non-queer tourists. Gay pride parades provide an opportunity to deconstruct acts of tourism, pleasure and politics as these are lived through the bodies involved.

Tourist activities and experiences are directly mediated through bodies. Sociologists Soile Veijola and Eeva Jokinen (1994: 149) emphasise that there is an 'absence of the body' in tourism studies and they pose the question: is it 'possible to thematise the embodiment, radical Otherness, multiplicity of differences, sex and sexuality in tourism?' (Veijola and Jokinen 1994: 129).

It is possible to respond to Veijola and Jokinen's (1994) question in two ways. First: to recognise that the study of tourism, within the social sciences, has been built on hierarchical dualisms and tends to produce hegemonic, disembodied, and masculinist knowledge. Furthermore, the mind and body are usually conceptualised as a dualism which is gendered and sexualised (Johnston 1996). Too often tourism research is presented as methodologically precise and statistically impeccable but otherwise disembodied. One of the central dichotomies in tourism studies is self/Other. Tourism can be theorised as the powerful manufacturer of the exotic (Rossel 1988) and a commodifier of cultures (Greenwood 1978) which constructs 'Others'. The relationship between power, meaning and knowledge within tourism has been given some attention in postmodern accounts of tourism (for example MacCannell 1992). A neglected part of the postmodern project in the social sciences, however, has been to acknowledge the importance of feminist and critical social theory on embodiment (Johnson 1989). Western philosophical traditions of dualities shape conceptualizations of both people and places (see Grosz 1992).

I begin this chapter by addressing the question 'what are queer bodies?' arguing that sexualised and gendered subjectivities are fluid, changing and at times paradoxical. Queer bodies, like all bodies, cannot be taken for granted or treated as obvious. The term 'queer' is a highly contested and political term that cannot be easily contained in a single neat dictionary definition. In order to understand queer bodies and tourism it is necessary to pay attention to paradoxical discourses of sexualities and genders. Second, I examine some of the ways in which gendered and sexualised bodies have been Othered in particular fields of tourism studies (including hallmark tourism, postmodern tourism, sex tourism and gender and tourism). My third point is that there has been an upsurge of interest in queer bodies and I outline the contested terrain of what it means to be, and write about, queer in the academy. Fourth, some of the recently emerging work that focuses on queer tourism is discussed, including research publications, special journals editions and conference sessions. In this section it is argued that although some tourism academics are now paying attention to queer bodies, it is still not acceptable for gendered queer materiality to be narrated. Finally, I examine particular aspects of queer theorising – notions of abjection, performativity and camp – and suggest that these are potentially useful concepts for social scientists who wish to dislodge the heteronormativity and hegemonic masculinity in tourism studies.

What is a queer body?

> Queer bodies may be understood as mismatches of sex, gender and desire.
>
> (Jagose 1996: 3)

Put another way, a queer body might be a body which unhinges the allegedly stable relationship between sex, gender and desire and 'exploits the incoherencies in those three terms which stabilise heterosexuality' (Jagose 1996: 3). Refusing to take heterosexuality as its starting point, queer bodies demonstrate the impossibility of any 'natural' sexuality and call into question usually taken-for-granted terms such as 'man' and 'woman'.

Bodies and their functions have long been perceived as 'natural' phenomena, as 'raw material', that is, as non- or pre-social. Yet 'the body' cannot be regarded as purely a social, cultural and signifying effect 'lacking in its own weighty materiality' (Grosz 1994: 21). Longhurst argues that while there has been an explosion of work on the body, 'the seemingly simple question "what is the body" has not been examined thoroughly' (Longhurst 2001b: 11).

Elizabeth Grosz (1992: 242) has been theorising gendered and sexualised embodiment for a number of years and she offers a definition of the body, claiming (italics in the original):

> By *body* I understand a concrete, material, animate organization of flesh, nerves, muscles, and skeletal structure which are given unity, cohesiveness, and organization only through their psychical and social inscription as the surface and raw materials of an integrated and cohesive totality. The body becomes a *human* body, a body which coincides with the 'shape' and space of a psyche, a body whose epidermic surface bounds a psychical unity, a body which thereby defines the limits of experience and subjectivity, in psychoanalytic terms through the invention of the (m)other, and ultimately, the Other or Symbolic order.

Grosz's definition sketches the possibilities of what a body might *be*, but the very matter of a body remains elusive and problematic.

Gendered/sexed and sexualised bodies are not fixed by nature, nor completely culturally constructed. 'The sexed body is not simply there, ready and waiting, for us to examine' (Cream 1995: 31). An analysis of gay pride parades cannot ignore the 'real' and exposed flesh which is explicitly gendered and sexualised.

The notion that all bodies fit into either a male or female has been hotly contested. Julia Cream (1995: 31) asks: 'What is the sexed body?' She insists that the sexed body is contingent upon time and place. Furthermore, the sexed body could be read as an historical outcome of a range of discourse and meanings centring on biological sex, social gender, gender identity and sexuality. Cream exposes the commonly held belief that we all have to fit into the model of one sex or another. There are many bodies, however, that do not fit the two sex model. Cream uses examples such as bodies that have both male and female genitalia, usually known as 'intersexed', transsexuals and people with XXY chromosomes.

Cream is one of many geographers who have become fascinated with embodiment (for a review of geographical work on bodies, see Longhurst (1997)) and 'the body' has become a useful resource to articulate, for example, a politics of difference. It is now well established that bodies and places are mutually constitutive (Nast and Pile 1998). 'The call to understand the body is often simultaneously a call for the fluidity of subjectivity, [and] for the instability of the binary of sexual difference' (Callard 1998: 388). Postmodernist metaphors such as mobility and fluidity are often used to describe bodies, places and their constitutive relationship. It is still not acceptable however, to speak of the actual *body fluids* that threaten to break the body's boundaries and mess up other bodies and places (Longhurst 2001a). Bodily fluids are not deemed a legitimate topic in studies of tourism because they threaten to mess up clean, clinical, statistical and hard (disembodied and masculinist) knowledge. When social scientists do speak of the body 'they still often fail to talk about a body that breaks its boundaries – urinates, bleeds, vomits, farts, engulfs tampons, objects of sexual desire, ejaculates and gives birth' (Longhurst 2001a: 23). The messy body is often Othered and feminised as a result. What constitutes appropriate issues and legitimate topics to teach and research comes to be defined in terms of reason, rationality and abstraction as though these can be separated out from passion, irrationality, messiness and embodied sensation (Longhurst 2001a).

Some social scientists theorising sexualised bodies, particularly those employing postmodern and critical social theories, tend to posit the body in the realm of ideas, rather than discuss the matter of bodies. Talking about the body is 'safer' than actually embodying our geographies when considering the reality of what are often homophobic disciplines. Jon Binnie (1997) notes that it is easier to theorise bodies, rather than discuss the actual materialities of gendered and sexualised bodies. He notes that the risk of not getting an academic job or promotion means that some work on sexuality and space is devoid of flesh. There is still squeamishness around sex and sexuality in the production of academic knowledge. While it may be acceptable to speak of sexuality, for example, it is the actual mention of embodied sexuality (the materiality of sex and sexuality) that creates disciplinary anxieties

(Bell 1995). This has certainly been the case when I have presented research on gender and sexuality. My research is occasionally mocked by colleagues and the popular press. For example, when in 1993 I undertook an Honours research project on the performance and surveillance of lesbian identities and domestic spaces one staff member did not think my work was 'geography' and wanted the research stopped. It is possible that the tolerance afforded to queer geographers, as noted by Binnie (1997), evaporates when confronted with the materiality of lesbian, gay, bisexual and transgendered bodies. In 1994 and 1995 I researched female body builders and gyms for a masters thesis. An editor for a national New Zealand newspaper printed a New Zealand Geographical Society advertisement for a seminar I was presenting and placed it under the title: 'It's getting worse folks' (*New Zealand Herald* 1995: 7). Attempts to belittle and marginalise University of Waikato feminist geographical work on gendered and sexualised embodiment has been documented elsewhere (see Peace, Longhurst and Johnston 1997). In 2000 at a UK University I presented research on mountains and masculinity (Morin, Longhurst and Johnston 2001) which was then joked about by some of my male colleagues in the departmental tea room. I offer these personal examples to illustrate that the connection between gendered and sexualised bodies and geography or tourism knowledges is often upheld as absurd, humorous and illegitimate.

Resistance to work on the body might stem from powerful modernist notions about the mind and body dualism. Dualistic thinking has been present throughout the history of western philosophy and has been the focus of many philosophers since Socrates (see, for example, Derrida 1981; Foucault 1970; and Nietzsche 1967, 1969).

> From the beginnings of philosophical thought, femaleness was symbolically associated with what Reason supposedly left behind ... maleness remained associated with a clear, determinate mode of thought, femaleness with the vague and indeterminate.
>
> (Lloyd 1993: 2–3)

Feminist philosopher Genevieve Lloyd (1993) examines the work of various philosophers (such as, Plato, Aristotle, Bacon, Philo, Augustine, Aquinas, Descartes, Hume, Rousseau, Kant, Hegel, Satre and de Beauvoir) in order to highlight that since the inception of philosophy as a discipline in Greece maleness has been associated with reason, the mind and abstraction, and femaleness with irrationality, the body and materiality. The early Greeks associated women with reproduction and nature but men were understood to be able to transcend their bodily functions (nature) because of their capacity for reason. Plato was reported as saying that women imitate the earth (Lloyd 1993). Lloyd traces this man/woman binary through to the sixth century BC when the Pythagoreans linked maleness to notions of enlightenment, orderliness and goodness. Femaleness was linked with darkness, disorderliness and badness. Explicit in this construction of opposites is the understanding that Woman is to be avoided as she is associated with negative and 'dangerous' qualities, while Man gains the highest status and is associated with virtuous qualities.

The determinate, rational world was aligned, in Plato's universe, with *form*, that is, the knowability of the objective world. The indeterminate was aligned with unknowable *matter* or nature. In early Greek philosophy, then, a bond between knowledge and rationality was established. This bond persists today, albeit in different forms.

These opposing positions gained more credibility during the Enlightenment and through Descartes's theory of separation between the knower and the known. Susan Bordo (1986) calls this the 'cartesian masculinisation of thought'. Bordo (1986: 450) argues that this separation 'made possible the complete intellectual transcendence of the body'. 'Masculinist rationality is a form of knowledge which assumes a knower who believes he can separate himself from his body, emotions, values, past experiences, and so on' (Longhurst 1995: 98). Larry Berg (1994: 249) adds: 'the mind was already associated with the masculine and the body was associated with the feminine, thus Descartes's mind/body dualism laid the conceptual groundwork for the masculine rational transcendence of the feminine irrational'.

Elizabeth Grosz (1989: xvi) argues: 'when the system of boundaries or divisions operates by means of the construction of binaries or pairs of opposed terms, these terms are not only mutually exclusive, but also mutually exhaustive' (Grosz 1989: xvi). It is important to highlight here that the two sides of the dualism are not unrelated. If one side is represented by 'A', then its opposite will be a conceptualisaion of what 'A' is not, say 'A-'. The sides of the dualism, therefore, have an epistemological relation. This is a mode of knowing in which A has a positive status and only exists in relation to its Other; 'the other term is purely negatively defined, and has no contours of its own; its limiting boundaries are those which define the positive term' (Grosz 1989: xvi). The classic examples, in relation to a study of gay pride parades, are that the terms 'Man' and 'heterosexual' have positive identities, while 'Woman' (or not-Man) and 'homosexuals' (or not-heterosexuals) have negative identities.

This 'crisis of reason' (Grosz 1993: 187) has forced social scientists to examine critically the production of knowledge and to question founding epistemological assumptions. The mind/body dualism has important associations with reason and masculinity. Grosz (1989: xiv) argues that the mind has been traditionally associated with positive terms such as 'reason, subject, consciousness, interiority and masculinism'. The body, however, has been negatively associated with 'passion, object, unconsciousness, exteriority, and feminism' (Grosz 1989: xiv). This western rationalist tradition entails a radical separation of mind and body that accords primacy to the mind. Lloyd (1993: 2) claims that 'from the beginnings of philosophical thought, femaleness was symbolically associated with what Reason supposedly left behind –

the dark powers of earth goddesses, immersion in unknown forces associated with mysterious female power'. It is necessary to 'examine the subordinated, negative or excluded term *body* as the *unacknowledged condition* of the dominant term, *reason*' (Grosz 1995: 32, italics in original). One strategy available to social scientists is to examine the role embodiment has to play in the production and evaluation of knowledges.

Veijola and Jokinen (1994) have begun a critical engagement with the gendered nature of the mind/body dualism within tourism studies. Some feminist geographers have deconstructed and reconstructed western binaries in an explicit poststructural, corporeal, feminist politics (for example Cream 1992, 1995; Johnson 1989; Johnston 1995, 1996, 1998; Longhurst 1995, 1997; Rose 1991, 1993a, 1993b). This work provides an important context within which to evaluate the construction of disembodied and masculinist knowledge. Longhurst (1995: 97) argues:

> a way for feminist geographers to subvert the hegemony that masculinity has over this knowledge may be to create an upheaval of the dominant/subordinate structure of the relation between the mind and body and to sexually embody geography.

Further, Longhurst (1995: 99) considers the body as geography's Other, claiming 'it has been denied and desired depending on the particular school of geographical thought under consideration.' Louise Johnson (1989: 134) argues 'instead of seeing the body as distinct from the mind, tied to a fixed essence or reduced to naturalistic explanations, it can be viewed as the primary object of social production and inscription.'

Despite these challenges, the mind/body dualism remains a dominant concept in western culture. We all have bodies, however, the difference is that men (who are assumed to be heterosexual, white, able bodied, of a particular social class and so on) are generally considered to be able to seek and explain pure knowledge, as if unencumbered by a material body. They attempt to transcend their embodiment by treating it as a mere container for pure consciousness. Women, homosexuals, blacks, disabled and so on, are thought to be bound to their fleshy bodies and unable to think straight as a consequence. One of the outcomes of this thinking is that queer bodies have been othered.

Othering bodies

Tourism knowledges tend to position bodies and places of the tourist gaze as exotic 'Others'. Tourism is constructed through difference, the exotic, recreation and displacement. John Urry (1990a: 11) argues 'Tourism results from a basic binary division between the ordinary–everyday and the extraordinary'. Thus, the study of tourism has come to be defined through hierarchical oppositions such as self/Other, tourist/host, same/different, work/play. These opposing terms are never neutral. The positive term is valued over and above the negative term. There is much potential to examine the gendered nature of these dichotomies. However, dominant discourses have constructed – for the most part – a masculine view of tourism as a product of waged labour classes in (post)industrial societies (Craik 1997). There are several assumptions built into this view. The most obvious example is that tourism research frequently leaves unexamined the differences of women's and men's experiences as tourists. The non-gendered and non-sexualised 'tourist' becomes masculine and heterosexual by default. If, instead, the gendered body is made explicit then the 'illusion that they [tourists] represent humanity in general is destroyed' (Jokinen and Veijola 1997: 36).

The tourism researcher has also been given universal status. There are important connections between western rationality and the right to know. It has been argued that 'an assumption of an objectivity untainted by any particular social position', or – one could argue – a specifically gendered/sexed body, allows masculinist rationality 'to claim itself as universal' (Rose 1993b: 7). Veijola and Jokinen (1994: 149: emphasis in the original) maintain that:

> judged by the *discursive postures* given to the *writing subject* of most of the analyses, the analyst himself has, likewise, lacked a body. Only the mind, free from bodily and social subjectivity, is presented as having been at work when analysing field experiences, which have taken place from the distance required by the so-called *scientific objectivity*.

The tourism studies analyst – what Donna Haraway (1991) would call, the master subject (in other words, the dominant subject constituted as white, bourgeois, masculine and heterosexual) – is omnipresent, both everywhere *and* nowhere to be found in tourism studies texts or the field.

While still remaining somewhat distant from their research, tourism analysts engage with the social effects on host communities surrounding events such as festivals, or hallmark events (see Hall 1992; Olds 1988; Roche 1990, 1991). These ensuing social impacts, however, tend to focus on employment, housing, infrastructurual changes and community identity. These studies begin to bring the Other into tourism research, but they also reconstitute the hierarchical relationships of dominant discourses. For example, Michael Hall (1992) discusses prostitution and crime as undesirable side effects of special events. Bodies involved in sex tourism become discursively constructed as deviant and are further Othered by the tourism researcher. Referring to a particular special event, Hall (1992) also discusses the 'hoon effect' of the Adelaide Grand Prix. The 'hoon effect' is, 'a reckless, irresponsible driver ... which may or may not have been encouraged by the staging of the Grand Prix' (Fisher, Hatch and Paix 1986: 152). It is important to note here that the driver, not surprisingly, remains non-gendered. If attention is paid to the masculinism of the hoon effect, then there is opportunity to make explicit dominant discourses of, for example, masculine-as-aggressive/feminine-as-passive. The marking-out and Othering of particular 'host' bodies in these tourism studies texts stands in contrast with the (usually) disembodied 'gazing' tourist.

Other studies of tourism phenomena have suggested that tourism must be seen as part of a postmodern valorisation of surface (see for example Cohen 1995; MacCannell 1992; Roszak 1986; and Selwyn 1990, 1996). Hence, tourism destinations are being theorised as fragmented collages of facts, clichés, 'nature', and history intertwined with entertainments of spectacle and carnival (Cloke and Perkins 1998).

Academic work on postmodernism and tourism has tended to use the phenomenon of tourism as a way of validating and celebrating difference and liminality. Tourism is often discussed in terms of creating marginal places and these 'liminal' places are privileged as sites of radical possibilities, away from the oppressive spaces/places of modernity. Places on the margins, however, can also be spaces of powerlessness where the place becomes more central than the bodies which exist in that place. Rob Shields' (1991) *Places on the Margin: Alternative Geographies of Modernity* is an example of a text that celebrates marginal places and Others the body in the process. Shields (1991: 73–116) discusses the British seaside resort of Brighton in a chapter entitled 'Ritual pleasures of a seaside resort: Liminality, carnivalesque, and dirty weekends'. For Shields (1991: 73), Brighton is a 'place on the margin' because of its reputation as a place for the (heterosexual) 'dirty weekend'. Binnie (1997), however, has charged Shields with being unaware that Brighton is also a particular liminal space for gay men, lesbians, *and* queer-bashers. Binnie (1997: 226) points out that:

> Shields makes little mention of Brighton's history of a safe haven for sexual dissidents – a retreat for lesbians and gay men (documented in Daring Hearts a collection of lesbian and gay life stories from Brighton in the 1950s and 1960s (Brighton Ourstory Project 1992)). Shields does however quote from tabloid newspapers describing the town as the 'AIDS Capital of Britain'.

Binnie (1997) argues that gay men are constructed as passive victims in Shields' narrative and inextricably linked to AIDS, thus rendering gay males as deviant. Gay men's bodies become mere containers or vectors for HIV/AIDS (see also Brown 1995). Clearly then, at work in Shields' (1991) postmodern liminal spaces of Brighton is the marginality and invisibility of the Other: gay male and lesbian bodies.

There are also a number of sociological authors who discuss postmodern tourism (especially Urry 1988, 1990a, 1990b, 1992), and others who have theorised post-tourism and post-tourists (Feifer 1985). The hyperreal world of the French semiotician, Jean Baudrillard (1988), inspired by his travels to

North America (see also Eco 1986), has established 'the quintessential postmodern tourism experience' (Munt 1994: 101).

Cultural meanings of shopping malls, theme parks, Disneyland, dockland regeneration and World Fairs have been theorised as fundamental to the restructuring of 'capitalism' and postmodern cultural shifts (see Featherstone 1991; Harvey 1989; Levine 1987; Shields 1988; Urry 1990a; Walker 1991). Ironically, the simulated environment of Disneyland is now being viewed as an essentialised and authentic 'American' cultural product. Erik Cohen (1995: 16) notes:

> Although created for commercial touristic purposes, Disneyland over time became an American cultural landmark. Despite its 'contrived' origins, it acquired a measure of 'authenticity' ... The analysis of the structure and the symbolism of Disneyland has disclosed its deep structural meaning in American culture.

In other words, the contrived postmodern landscape of Disneyland has acquired recognisable and dominant meanings which can be read by tourists.

The common thread in these writings on postmodern tourism is that of simulated environments. Sharon Zukin (1991) and Ed Soja (1992) are geographers working with 'hyperreal' spaces. Derek Gregory (1994: 157–59) critiques their texts for being disembodied and argues that: 'in odysseys through postmodern spaces and over postmodern landscapes they [Zukin and Soja] have also – and less accountably – lost sight of Lefebvre's defiant insistence on the body as the site of resistance.' This is another example of a focus on place and the denial of bodies. There are other binary divisions emerging in tourism studies that work to produce disembodied knowledge, for example the gender/sex division.

The term gender is historically specific. Gender was developed in contrast to the term sex, to depict that which is socially constructed as opposed to biologically given. Gender and sex were understood to be distinct. In 1968, psychologist Robert Stoller published a book called *Sex and Gender*, in which he argued that a person's gender identity is primarily the result of psychological influences. The biological sex of a person signifies, but does not determine, the appropriate gender identity for that person.

In the 1970s feminists began to utilise the gender/sex distinction to make political ground. Women began to argue persuasively that gender was a culturally constructed notion that varied across time and place. The introduction of the gender/sex distinction was seen as a political attempt to intervene into the western world that declared women as 'different' because of their biology, their sex. It held out the 'promise of enabling an analysis of male privilege as the product of historically and culturally constituted systems of gender inequality, not as the natural outcome of biological differences between males and females' (Yanagisako and Collier 1990: 131). The gender/sex distinction is murky. 'It is not clear how one can eliminate the effects of (social) gender to see the contributions of (biological) sex' (Grosz 1994: 18). This bodily uncertainty can create political opportunities for destablising the mind/body dualism. The use of the term gender/sex can indicate the tensions and possibilities between gender and sex, or mind and body.

Some feminist theorists in tourism studies have argued that dualisms such as private/public and home/abroad are gendered (Enloe 1989; Morris 1988a, 1988b; Wolff 1995). There has been discussion of the body images that are frequently used to sell holidays, tourist destinations and tourist events. Travel brochures ironically represent prospective holiday makers in idealised settings: the sun is golden, the sea is sparkling, the sky is blue and the tourists semi-naked, bronzed, and relaxed (Marshment 1997). Bodies of scantily clad 'natives' are suggestive of exotic places and people. Cynthia Enloe (1989: 28) argues that the desire to know another place is conflated, in the tourist imagination, with women 'as the quintessence of the exotic ... something to be experienced'. The dominant position of spectatorship has been a masculine one (see Mulvey 1975, and Doane 1990), irrespective of the gender/sex of the spectator. The dominant tourist imagination could be understood as reinforcing the connections between the feminine and the body. For example:

> Predominant tourism brochure representations of men [are] associated with action, power, and ownership, while women are associated with passivity, availability, and being owned. From this perspective, uses of women, sexual imagery, and exotic markers in the tourism industry to market destinations are seen to often reinforce stereotypes and hierarchical divisions of labour. Host societies differentiated by race/ethnicity, colonial past, or social position from the consumer societies are sold feminised images.
>
> (Swain 1995: 249)

Implicit in this description is the connection between masculinity and the mind and between femininity and the body.

The topic of 'gender in tourism studies' is currently gaining wider readership. For example, the *Annals of Tourism Research: A Social Science Journal* (1995) devotes a special issue to gender. Vivian Kinnaird and Derek Hall (1994) engage in issues of gender and tourism but tend to use the term gender as an understood and undefined category. What is notable about these works is their concentration on the social construction of gender, in ways which can reinforce a division between sex and gender. Margaret Swain (1995: 247) in her introduction to the 'Gender in Tourism' special issue, strengthens this dichotomy when she argues: 'gender is, therefore, quite distinct from biological differences between the sexes, and is the basis of both women's subordination and potential change toward equality between women and men.' Swain (1995) refers to Veijola and Jokinen's (1994) work on the body and tourism but reasserts the gender/sex dichotomy when she states:

> Their [Veijola and Jokinen] culturally constructed socially contained femininity raises the questions they pose to their fictionalised companions on a Finnish style vacation to Mallorca. For Veijola and Jokinen 'the body' is emblematic of what is missing in universalizing social sciences theories in general and in tourism studies specifically.
>
> (Swain 1995: 258)

By relegating the material body to representation (emblem), Swain (1995) omits the flesh and reasserts gender as a social construction. Swain (1995: 258–259) instead focuses on matter-free 'ideologies of masculinity and femininity in host and guest populations; and the social divisions of labor, power and sexuality.'

While this work is an extremely important starting point, it rarely challenges the foundations of tourism studies. It is no longer adequate to add gender to an unexamined and unaltered foundation. These efforts in tourism studies have unwittingly repeated the privileging of the mind over the body thereby raising other problems.

Trying to overcome hegemonic knowledge practices, Swain (2004) reconsiders her own body in an article entitled '(Dis)embodied experiences and power dynamics in tourism research'. Swain (2004) draws on some of her work on gender and ethnicity issues for women producers of ethnic arts to show that she has come to occupy many positions as a researcher. Her white, female, straight, middle-class American body is crucial to the production of tourism knowledge.

Homosexualities, gay or queer bodies, however, have largely been excluded from tourism studies' discourse. Much of what has been written tends to define gay tourism through economic possibilities (the pink pound/dollar), thus further Othering and denying queer bodies (Holcomb and Luongo 1996; Torres-Kitamura 1997; and *Time* Magazine 1995). There has been some attention paid to gay (male) identity and the spaces of tourism (see Hughes 1997, 1998), and there has been a proliferation of work on the themes of sexuality and space (see, for example, Bell and Valentine 1995; Binnie 1997; Namaste 1996). Elsewhere, however, 'homosexuals' appear unproblematically in the sex tourism and HIV/AIDS literature (Cohen 1988). Chris Ryan and Michael Hall's (2001) discussion of sex tourism is a useful example of the trend by tourism scholars to foreground sexuality via the materiality of sexual practices, and thereby collapsing sex and sexuality. Ryan and Hall (2001) and Stephen Clift and Simon Carter (2000) discuss gay and lesbian tourism, but do not engage in the gendered dimensions of sexuality and tourism.

Annette Pritchard, Nigel Morgan, Diane Sedgely and Andrew Jenkins

(1998: 273–282) have identified a 'gay gap' in tourism literature and published an article entitled 'Reaching out to the gay tourist: Opportunities and threats in an emerging market segment'. They attempt to raise 'some issues surrounding gay tourism, gay destinations and gay space' (Pritchard *et al.* 1998: 280). The authors acknowledge the heterosexuality of public spaces but tend to essentialise 'gay' identity as a singular and static subjectivity. Their article is directed at the 'tourist industry' and focuses mainly on economic factors, for instance:

> It seems likely that more and more marketers will seek to court the gay consumer. Whilst actively targeting the gay market as potential customers may not be relevant or sensible for all in the tourism industry, it will become increasingly important for it to avoid negative stereotypes and unwitting offensiveness. As more and more gay people come out, the average consumer is increasingly likely to know a gay man or lesbian woman.
>
> (Pritchard *et al.* 1998: 280)

There are several hegemonic assumptions in this advice for tourism marketers. The average consumer here is constructed as heterosexual. A 'gay man' or a 'lesbian woman', however, is marked as the Other. The binary between hetero/homo remains fixed. Pritchard *et al.* (1998) draw on recent spatial work on sexualities published in *BodySpace: Destabilizing Geographies of Gender and Sexualities* (in particular, Chouinard and Grant 1996; Duncan 1996; Myslik 1996; Valentine 1996); however, closer attention to the ways in which space constructs sexualised and gendered bodies would strengthen their argument. Furthermore, Puar (2002c: 936) notes, in relation to Pritchard *et al.* (1998) that: 'it is one thing to state that heterosexuality is assumed in space, and quite another to proclaim that space is heterosexual'. If tourism studies' academics wish to 'materialise' gay and lesbian tourism, it must be balanced by a critical reflection on what the bodily identities 'gay and lesbian' mean, especially in particular places.

Tourism studies privileges gender as a social construction, as opposed to sex, which is seen to be biologically fixed. This implies: 'the omission of the body as a vital element in the constitution of masculine and feminine identity' (Johnson 1989: 18). Moreover, Johnson (1989: 18) claims that feminists: 'in their zeal to avoid the accusation of biologism and by embracing the logics of historical materialism and liberalism, have ignored the possibilities of examining the sexed body in space.' The gendered/sexed and sexualised body becomes denied and Othered in tourism studies' accounts of gender.

An examination of gendered embodiment at Pride needs to take into account the historical construction of heterosexuality and expressions of resistance to dominant discourses. Rather than conceptualise homosexuality and heterosexuality as distinct and separate, another approach is to position homosexuality and heterosexuality as thoroughly entwined (Foucault 1981).

This approach insists that heterosexuality depends on homosexuality. Historically, heterosexuality is understood as being derived from homosexuality. According to Foucault (1981) it was not until 1870 that homosexuality was specified as a particular identity category and the concept of heterosexuality could consequently emerge. Heterosexuality, therefore, requires a material conception of homosexuality and the active Othering of homosexuality in order to define itself as a distinct social form (Butler 1990).

Queer tourism

In recent years academics have been paying attention to queer tourism. Sociologists Soile Veijola and Eeva Jokinen were, to my knowledge, the first to raise concerns regarding gendered and sexualised bodies and tourism when they posed the question: is it 'possible to thematise the embodiment, radical Otherness, multiplicity of differences, sex and sexuality in tourism?' (Veijola and Jokinen 1994: 129). The materiality of queer bodies has not escaped Veijola and Jokinen (1994) and they argue for the inclusion of the body in tourism studies. Constructing imagined and daring dialogues (but relying on 'real' quotes), they critique various tourism theoreticians as producers of disembodied tourism studies' knowledge. (For example, Kippendorf 1987, MacCannell 1989, Rojek 1993, and Urry 1990a.) Veijola and Jokinen (1994: 149) argue that: 'the tourist has lacked a body because the analyses have tended to concentrate on the gaze and/or structures and dynamics of waged labour societies'. They draw on the work of feminist poststructuralist Judith Butler (1990) and Ann Game (1991) to identify the importance of gendered/sexed embodiment to the study of tourism. Veijola and Jokinen (1994) bring into tension the gender/sex and social constructionism/essentialism dichotomies by beginning with the 'real' body. For example, at one point in their tourist journey at the beach they remark:

> At that moment, a wet and gritty ball lands in the middle of us, followed by fluent international apologies. I turn around, annoyed, to inform the intruder about the unwritten rules on public beaches, but fall silent again when facing a naked man – or to be precise, his sex, which is *not* socially constructed.
> (Veijola and Jokinen 1994: 140, emphasis in original)

Also discussing the materiality of bodies is feminist philosopher, Vicky Kirby (1997: 2–4). She challenges the matter of the body as universally taken for granted in western knowledge and so begins a series of what she calls 'dumb questions' about what the body might be. Kirby (1997) uses the Hindu festival ritual of *thaipusam* to illustrate that the matter of the body is not straight forward. Tourists at this ritual become astonished voyeurs at the spectacle of a man's body, grotesquely impaled on elaborate metal spokes which are driven into the skin and organs of his body. His hands, face, lips and neck are also skewered with long spikes. Kirby (1997: 3) notes:

> this man does not bleed, nor does he scar ... However, this cultural/ritual incorporation is not generalizable, for it does not extend to the bodies of tourists, or even other members of the devotee's own community who might witness the festival.

Kirby (1997: 3) suggests that the cultural context that surrounds a body can also come to inhabit it. At gay pride parades, the paraders are impaled, in a sense, by the tourist's gaze. Unlike the religious rituals described by Kirby (1997), a pride parade – usually – has the purpose of secular display. It may not, however, be any more accessible to the watching tourists than an extreme religious ritual. The cultural context of 'queerness' surrounds and arguably inhabits the paraders, but to some extent (as discussed later in Chapters 3 and 4) also extends to the bodies of the watching tourists.

Despite Kirby's (1997) claim that ritual incorporation is not generalisable, bodies involved in the tourism process do undergo change. Quetzil Castaneda (1991: 216 cited in Veijola and Jokinen 1994: 147) argues that:

> the tourist attempts to redefine their body in contrast to the Other's Body and to redefine it in order to attach it with the category of Self that is being upheld and which can only be seen through the reflexive play of the Other, as a category and as Body.

This explicit attention to gendered and sexualised embodiment offers many possibilities for tourism studies. Tourism studies research which includes the Other (that is those people defined as homosexual, poor, black, diseased, working class and so on) is an important starting point for retheorising tourism studies. The next step, however, is for tourism studies research to focus on unsettling the dominant/subordinate structure between mind and body, and between heterosexual and homosexual, self and Other, thereby challenging the masculinism of tourism studies. Focusing attention on the gendered/sexed and sexualised bodies of gay pride parades can prompt new understandings of power, knowledge and social relationships between bodies and tourism processes.

Despite these risks associated with research on gendered and sexualised bodies many geographers are focusing their studies on embodied performance, bodily displays and the ways in which bodies act out multiple and fluid sexual identities (see the theme issue 'Spaces of Performance, part one and two' in *Environment and Planning D: Society and Space*, 2000). There is some discussion surrounding the construction of gender and race at street festivals and carnival events (see Jackson 1988, 1992; Lewis and Pile 1996; and Spooner 1996). Notably, Clare Lewis and Steve Pile (1996) offer complex readings of women's bodies at the Rio Carnival which renders bodies and identities indeterminable.

In 1997 I attended a conference entitled 'Gender/Tourism/Fun?' at UC Davis. This conference brought together academics who were critical of the exclusion of women and the analysis of gender in tourism knowledges. There were some exceptions. The most obvious was the keynote speaker – Dean MacCannell – who failed to mention or theorise gender. Noticeably, too, throughout the conference was the absence of any discussion of homosexuality and tourism and, perhaps more concerning, a lack of engagement with heteronormative tourist acts and processes.

A special issue of the journal *Gay and Lesbian Quarterly* (*GLQ*) (2002) entitled 'Queer Tourism: Geographies of Globalization' is the first edited collection to address same-sex sexuality and travel. The special issue documents, describes and theorises the growth of gay and lesbian tourism as well as other forms of queer travel. Furthermore, the special issue makes explicit the importance of queer sexualities to tourism, globalisation, and economic systems. Editor Jasbir Puar (2002a: 1) notes: 'queer tourism is still one of the least researched or discussed topics in scholarly venues'. The collection reflects interdisciplinary discussions that are informed by queer and feminist theory, ethnography and globalisation studies. Many of the articles focus on spatial politics of tourism, such as Venetia Kantsa's (2002) account of lesbian tourism in Eresos, on the Greek island of Lesvos. Kantsa (2002) documents the bodily and landscape changes of the lesbian resort and argues that this is an extraordinary place that enables lesbian gazes, a focus on the history of the lyric poet Sappho, and women-only spaces that are usually not feasible in other tourist places. Puar (2002a: 3), who is dismayed at the lack of attention in queer tourism paid to women, lesbians and gender, notes that Kantsa's article is unique in that it is 'possibly the only to focus exclusively on lesbian and queer women's tourism'. While adding analyses of lesbians to tourism studies begins to challenge tourism's masculinity, it may be more productive to consider, however, the ways in which space, mobility and tourism are gendered and sexualised. Puar (2002a: 4) states: '[T]he focus on men's travel, to the exclusion of women's, is both a historically entrenched problem and a failure to incorporate gendered analyses into conceptualisations of tourism and travel.'

Other papers in this collection begin to question and theorise the relationship between queer tourism and globalisation. Gabriel Giorgi (2002), in his article 'Madrid En Tránsito: Travelers, Visibility, and Gay Identity' argues that the political transformations in Spain have made gay and lesbian tourism in Madrid possible. He argues that gay tourism becomes a sign of global circulation and that it 'sets in motion a narrative that locates bodies in a geopolitical order, making them visible in some ways and determining their visibility under different conditions' (Giorgi 2002: 73). Also discussing cities and queer visibility, Dereka Rushbrook (2002) argues that local changes in queer space are related to the global commodification of space and bodies through tourism. She found cities' queer spaces came to function – and were promoted – as exotic mixes of the Other. 'The promotion of gay neighborhood as yet another commodity leads to a form of assimilation into mainstream culture that reinforces the assimilation created by the production of the gay and lesbian niche market' (Rushbrook 2002: 194). She found that entrepreneurial states actively promote a cosmopolitan identity in order to attract tourists and as a result queer spaces are then marketed for and touristed by non-queers.

State practices are the topic of Puar's (2002b, 101–137) article in this collection entitled: 'Circuits of Queer Mobility: Tourism, Travel and Globalization'. She documents several shifts in the gay and lesbian tourism industry during the last ten years which include an increase of international travel, a change from 'private' industry's marketing to the nation's 'public' marketing aimed at queer travellers. She is critical of the ways in which the tourism industry reifies queer travellers as gay white affluent males and calls for a more critical approach to the study of queer tourism that would include different axes of subjectivity, such as gender, race, class and so on.

Two more papers in this special issue of the *Gay and Lesbian Quarterly* directly address gay pride parades. One is an ethnographic account by a journalist at World Pride Roma, 2000 (Luongo 2002) and I draw on this work in more detail in Chapter 6. The other paper engages with material from the Sydney Gay and Lesbian Mardi Gras (Markwell 2002). Both of these articles offer insights into how global tourism has altered the spatial relations of place and subjectivities of local regions. Markwell (2002) highlights the relationship of place to time and notes that Sydney Mardi Gras is successful because of 'temporal containment' that ensures the transgressive features are moderated by the limitations of place and time. As a consequence, particular bodies dominate the event, while other queer bodies, such as working class and non-white lesbians and gays, are marginalised.

In the remainder of this book I build on this aforementioned work to discuss the queer spaces of bodies and tourism at pride parades. I draw on a variety of theories from a 'grab bag' of queer theorising. In the following section I outline notions of abjection, camp and performativity. These theories provide a basis for the empirical material that is loosely nested into different spatial scales: bodies (Chapter 3), streets (Chapter 4), suburbs (Chapter 5) and cities (Chapter 6).

Critical queer theorising

Queer theory is a term coined by Teresa de Lauretis and first appeared in her introduction to the special issue of *differences* called 'Queer Theory: Lesbian and Gay Sexualities' (de Lauretis 1991). She traces the way the conceptual framework for analysing sexualities shifted in the twentieth century from an early emphasis on 'homosexuality' to a gender undifferentiated 'gay', to more recently, 'lesbian and gay'. Concerned with the maintenance of differences, she remarks that specificity is elided 'in the contexts in which the phrase [gay and lesbian] is used, that is to say, differences are implied [in the phrase] but then simply taken for granted and even covered over by the word ' "and" ' (de Lauretis 1991: v–vi). The turn to the word 'queer' is meant to mark 'difference between and within lesbians, and between and within gay men, in relation to race and its attendant differences of class or ethnic culture, generational, geographical, and socio-political location' as central (de Lauretis 1991: viii).

> Queer theorists view heterosexuality and homosexuality not simply as identities or social statuses but as categories of knowledge, a language which frames that which we know as bodies, desires, sexualities, identities. This is a normative language as it shapes moral boundaries and political hierarchies. . . . Queer theory is the suggestion that the study of homosexuality should not be a study of a minority – the making of the lesbian/gay/bisexual subject – but a study of those knowledges and social practices that organise 'society' as a whole by sexualizing – heterosexuality or homosexuality – bodies, desires, acts, identities, social relations, knowledges, culture, and social institutions.
> (Seidman 1996: 12–13)

Queer theory provides a framework from which to draw on critical social theories which challenge heteronormative discourses. In particular, I draw on three specific, but overlapping, theories in this book: abjection, camp and performativity.

Abjection

Julia Kristeva's (1982) notion of abjection is very useful in understanding the ways heteronormative tourists at gay pride parades combine feelings of fascination and revulsion towards queer bodies. According to Kristeva the feeling of abjection is one of disgust, often evoking nausea, and it is:

> an extremely strong feeling which is at once somatic and symbolic, and which is above all a revolt of the person against an external menace from which one has the impression that it is not only an external menace but that it may menace us from the inside. So it is a desire for separation, for becoming autonomous and also the feeling of impossibility of doing so.
> (Kristeva 1982: 135)

That which is abject is something so repulsive that it both attracts and repels; it is both fascinating and disgusting. The abject exists on the border, but does not respect the border. It is 'ambiguous', 'inbetween', 'composite' (Kristeva 1982: 4). The abject is what threatens identity. It is neither good nor evil, subject nor object, but something that threatens the distinctions themselves. Kelly Oliver (1993: 56) claims: 'Every society is founded on the abject – constructing boundaries and jettisoning the antisocial – every society may have its own abject.' Kristeva (1982) maintains that the impure can never be completely removed. Abjection's ambiguity means that while releasing a hold, it does not radically cut off the subject from what threatens it; on the contrary, abjection acknowledges it to be in perpetual danger. David Sibley (1995: 8) adopts Kristeva's notion of abjection to argue that:

> the urge to make separations, between clean and dirty, ordered and disordered, 'us' and 'them', that is, to expel the abject, is encouraged in western cultures, creating feelings of anxiety because such separations can never be finally achieved.

'Us and them' discourses are evident at pride parades and may be predicated on ideas of abject objects, states, zones, agents, groups, psychic and political processes, as Anne McClintock (1995: 72; emphasis in original) outlines:

> With respect to abjection, distinctions can be made, for example, between abject *objects* (the clitoris, domestic dirt, menstrual blood) and abject *states* (bulimia, the masturbatory imagination, hysteria), which are not the same as abject *zones* (the Israeli Occupied Territories, prisons, battered women's shelters). Socially appointed *agents* of abjection (solders, domestic workers, nurses) are not the same as socially abject *groups* (prostitutes, Palestinians, lesbians). *Psychic* processes of abjection (fetishism, disavowal, the uncanny) are not the same as *political* processes of abjection (ethnic genocide, mass removals, prostitute 'clean ups').

'Socially abject groups' are the topic of Iris Marion Young's (1990: 142) discussion of what she called 'ugly bodies'. She argues that abjection enhances 'an understanding of a body aesthetic that defines some groups as ugly or fearsome and produces adverse reactions in relation to members of those groups' (I. M. Young 1990a: 142). This work is particularly useful when it comes to understanding why some people are both drawn to and repulsed by queer bodies in pride parades. Young (1990: 142) states that: 'racism, sexism, homophobia, ageism and ableism are partly structured by abjection, an involuntary, unconscious judgement of ugliness and loathing'. Young (1990: 146) goes on to argue that: 'homophobia is one of the deepest fears of difference precisely because the border between gay and straight is constructed as the most permeable; anyone at all can become gay'. The border is most threatening when the gay body cannot be distinguished from the straight body. Only when gay bodies are clearly marked as different, as in gay pride parades, does this border become visible and therefore less threatening to dominant groups.

Abjection, then, may account for the disgust and repulsion expressed at pride parades. I turn to another aspect of queer theorising – camp – in order to discuss strategies that queer bodies may employ when being an abject social group weighs too heavily.

Camp

Binnie (1997) notes that the notion of a single observable truth has been used to silence and marginalise lesbian and gay bodies and places. He goes on to state that: 'camp can productively work to undermine accepted values and truth, especially the heterosexual definition of space' (Binne 1997: 229). Possibilities of camp are easier to document than to define because they tend to defy discrete, historical and aesthetic categories. 'Representational excess, heterogeneity and gratuitousness of reference ... signal and contribute to an overall resistance to definition' (Cleto 1999: 3). One of the possibilities of camp is that it can facilitate the creation of non-heteronormative bodies and spaces in oppressive locations. Richard Dyer (1993: 135) states that: 'all the images and words of the society express and confirm the rightness of heterosexuality. Camp is one thing that expresses and confirms being a gay man'. Camp can, I believe, be a useful theoretical possibility to dismantle the static constructions of masculine/feminine and heterosexual/homosexual. This is evident in Butler's (1990: 31) discussion of the campness and subversiveness of gay drag (italics and underlining in original):

> The replication of heterosexual constructs in non-heterosexual frames brings into relief the utterly constructed status of the so-called heterosexual original. Thus, gay is to straight *not* as copy is to original, but, rather, as copy is to copy. The parodic repetition of 'the original' ..., reveals the original to be nothing other than a parody of the idea of the natural and the original.

The employment of a camp sensibility – a type of appearance and behaviour that mocks and plays with gender norms – undermines gender assumptions and can create an upheaval of dualisms such as man/woman and heterosexuality/homosexuality in particular cultural contexts. There has been some criticism regarding the notion of camp. While a certain type of 'togetherness' can be gained from being camp, Dyer notes (2002) that it has some drawbacks. There tends to be 'an attitude that you can't take anything seriously, everything has to be turned into a witticism or joke' (Dyer 2002: 50). Furthermore, self-mockery as self-protection can begin to erode queer subjectivities and gays and lesbians may begin to internalise that which is mocked. Progressive or reactionary, the practice of camp is ambiguous. Parades as tourism events may be illuminated by notions of camp because when it comes to camp, context is important. Camp works when things – bodies, places and times – are relational. Camp sensibility then, is very much a product of, and a type of resistance to, oppression.

Performativity

It is possible to make further trouble for heteronormative discourses by drawing on Butler's notion of performativity. Butler's (1990: 33) now well known theory of gender, is: 'the repeated stylisation of the body, a set of repeated acts within a highly rigid frame that congeal over time to produce the appearance of substance, of a natural sort of being'. She (1990: 136) discusses the performative as (italics in original):

> ... acts, gestures, enactment, generally construed, are *performative* in the sense that the essence or identity that they otherwise purport to express are *fabrications* manufactured and sustained through corporeal signs and other discursive means. That the gendered body is performative suggests that it has no ontological status apart from the various acts which constitute its reality.

I argue that the act of parade touring further exaggerates the instability of heteronormative discourses which in turn produces fabrications of subjectivities. In *Environment and Planning D: Society and Space* special issue on performativity, Nicky Gregson and Gillian Rose (2000: 434) argue that:

> Performance, in short, seems to offer intriguing possibilities for thinking about the constructedness of identity, subjectivity, and agency.... [A] notion of performance is indeed crucial for a critical human geography concerned to understand the construction of social identity, social difference, and social power relations, and the way space may articulate all of these.

I agree with Gregson and Rose (2000) that performance and performativity are important conceptual tools for critical geographers and others concerned to denaturalise taken for granted social practices and norms. Gay pride parades involve a saturation of performances and performers of power, with particular subject positions. In other words, various subject positions – paraders, tourists, sponsors, and so on – become produced by power. The performances of subject positions are iterative and therein lies the potential for disruption, as there is no guarantee that its repetition will be successful. With this in mind, I not only highlight the constructed hegemonic norms, but I also look for the cracks and slippages that occur in the performativity of parades. Of particular interest in this project are the implicit and normalised discourses of heterosexualised bodies, and their associated constructions of gender and gendered roles, that surround understandings of tourism, bodies, streets, suburbs and cities.

Geographers have taken up this notion of camp sensibility and the arguments put forward by Butler (1990) to think about the performance of sexual identities in space. David Bell, Jon Binnie, Julia Cream and Gill

Valentine (1994: 31) explore the ways in which queer identities are oppositional, transgressing and parodying heteronormativity through a 'long hard look at two current dissent sexual identities – the hypermasculine skinhead and the hyperfeminine lipstick lesbian'. Bell *et al.* (1994: 33) make important contributions that trouble gender and upset 'straight' spaces, claiming that the 'mimicry of heterosexuality by gay men and lesbians has the potential to transform radically the stability of masculinity and femininity, undermining its claim to originality and naturalness'. The authors are critical of Butler's work and cite Susan Bordo (1992: 171) who points out that Butler 'does not consider the possibility of different responses of different "readers" (male/female, black/white, young/old, gay/straight, etc.) or the various anxieties that might complicate their readings'. Furthermore, Bordo (1992: 171) states: 'when we attempt to give [Butler's] abstract text some more "body" we run into difficulties'.

In this book I include multiple audience responses and focus on the difference that space makes in order to provide a link between theory and empiricism. While Bell *et al.* (1994) offer detailed case studies of gay skinheads and lipstick lesbians, there is room to theorise further the spaces that constitute these identities. Their article prompted four responses (Kirby 1995; Knopp 1995; Probyn 1995; Walker 1995), in which Probyn (1995: 81) urges us to remember 'that the conditions of the production of space as gendered or as sexed are historically, materially and strategically different'. Probyn offers a description of two women in the masculine space of a pub to illustrate the difference that gender and sexuality make and notes that the lesbian subject is always a doubled subject, caught up in being a woman and a lesbian.

In the chapters that follow I draw on the notion of abjection, camp and performativity. First, I focus on the gendered bodies of pride parades. In particular, I take a long hard look at marching groups – male and female – in order to understand how their mobile subjectivities come into being in relation to tourism, tourists, and each other. Camp and performativity theories guide Chapter 3. Second, I try to make sense of some abject reactions to queer bodies in parades (Chapter 4) by focusing on the borders that are maintained and broken on the streets of pride parades. Third, I consider parade site politics where a long and vocal debate raged regarding the relocation of the HERO Parade from Auckland's CBD to a 'gay suburb' (Chapter 5). Finally, I focus on city tourism and commodification to unravel the paradoxes of pride events that are about protest and entertainment (Chapter 6). These chapters – bodies, streets, suburbs and cities – illustrate the potential for tourism researchers to engage with queer theories about specific bodies in particular material spaces.

3 Bodies
Camped up performances

Queer bodies in gay pride parades may 'camp-up' heteronormative notions of what it means to be gendered and sexualised. In this chapter I follow Jon Binnie's (1997: 223) direction 'towards a queer epistemology' and adopt a camp appreciation for a particular type of pride performance – marching groups. Male and female marching groups are integral to the success of Pride and they highlight the embodied constructedness of gender and sexuality.

Butler's (1993) notion of performativity has moved social theorising away from essentialised notions of gender and sexuality. An embodied analysis of Pride matters because it represents a site through which marginalised groups attempt to challenge heteronormative tourist spaces. Notions of camp are employed because they trouble that which is considered natural and highlight relationships between power, knowledge, subjectivities and spaces. Two very different marching group performances – a women's drumming group in Pride Scotland, Edinburgh, and male marching groups in Sydney and Auckland parades – are compared in order to argue that space makes a difference to the enactment and embodiment of camp.

Bell *et al.* (1994: 31–48) discuss the ways sexual identity is performed in space. They describe the hypermasculine 'gay skinhead' and the hyperfeminine 'lipstick lesbian' and assess the effects of their performance in spaces which are actively constructed as heterosexual. Bell *et al.* (1994) make important contributions that trouble gender and upset 'straight' spaces. They provide some points of connection and departure in my project for untangling the ways in which bodies are gendered and sexualised at Pride.

Drawing on Grosz (1992) I argue that bodies and places are mutually constituted and hence bodies are gendered and sexualised according to particular times and places. Space is bound into power/knowledge relationships and, therefore, the parades as tourist sites/sights are central to the subjectivity of the paraders. Rose (1997) asserts that 'particular imagined spatialities are constitutive of specific subjectivities. Identities are in part constituted by the kind of space through which they imagine themselves.'

Grosz (1994: 142) pushes the constitutive relationship between bodies and environments further, arguing that:

it is crucial to note that these corporeal inscriptions do not simply adorn or add to a body that is basically given through biology; they help constitute the very biological organization of the subject – the subject's height, weight, coloring, even eye color, are constituted as such by a constitutive interweaving of genetic and environmental factors.

Bodies, therefore, do more than 'perform' or 'put on' a particular gender and sexuality. Bodies are gendered in a way that can challenge the very *matter* of bodies.

Marching bodies at Pride are caught up in, and constituted by, other (institutional) biopower discourses relating to the gendered and sexualised bodies of gay pride parades. When these built bodies use power through the medium of their bodies, certain 'contradictions' emerge. Responses from tourists indicate that the process of constructing sexuality is highly dependent on the constitutive relationship between paraders and tourists. The majority of tourists reinscribe themselves as 'normal' and/or heterosexual, despite a variety of subject positions that are on 'offer' in the parading carnivalesque spaces. City policies which attempt to control gay pride parades illustrate unease with gay bodies in 'everyday' city streets.

A collection of papers which begin to question the relationship between sexuality and gender as well as sexuality and race appear in *Antipode* under the special issue title 'Queer Patriarchies, Queer Racisms, International' (2002). This is a timely collection for geographers as little has been documented or published on tensions that exist between scholars working in the field of sexuality. The collection addresses the 'crosscurrents of patriarchies, racisms, and classisms that flow across sexuality formations and sexed places and in so doing traces out critical lines of sociality and communication amongst all sorts of queer and not-so-queer lives' (Nast 2002: 841). In a sense, this collection muddies the waters for academics who research queer subjectivities and spaces. The contributors to this collection refuse assumptions that same-sex sexuality is necessarily liberating by virtue of its practice, nor necessarily anti-hegemonic normative sexuality. It questions – as it highlights – the inequalities that exist within sexualised subjectivities.

Heidi Nast (2002: 835) highlights what she calls 'palpable, gendered tensions among queer folk' at several Sexuality and Space Specialty Group (SSSG) sessions of the Association of American Geographers (AAG). Her (2002) contribution to this debate is to discuss the discursive construction of queer white patriarchy through an examination of magazines, advertisements, and tourist resorts in the US, and to raise concerns about gay men's access to better education, higher paying jobs, less child and elderly care responsibilities. Nast's (2002: 836) central question is whether or not certain gay white men are necessarily less patriarchal or racist because they are gay, which raises 'larger questions, about how (and in what) all queer lives are invested'. Like Nast, Virginia Blum (2002), in the same collection, articulates the ways in which queer subjects form their relationships in the context of consumer capitalism.

The absence of work in tourism studies dealing with women, lesbians and gender is highlighted by Jasbir Puar (2002c). Puar (2002c) thinks through the relationship between queer tourism and space using theories of intersectionality. She maps out the ways in which queer tourism and queer space occlude questions of gender.

Following Puar's (2002c) call to consider gender in queer tourism, the next section focuses on the make up and performances of male and female marching groups. Male marching groups – encouraged by watching tourists – camp up their performances to exaggerate notions of masculinity and femininity. The female marching groups under consideration in this chapter use camp sensibilities to dispel tourists' hostile reactions. I wish to point to the ways these marching groups queer understandings of what is feminine and/or masculine. Through a critical examination of the bodies of the marching groups, I suggest a number of deconstructive readings of hegemonic notions of gender and sexuality which enables a rewriting of gendered and sexualised bodies in ways that may influence mobile acts of politics and tourism.

I draw from empirical material collected at the following gay pride parades: Aotearoa/New Zealand, Auckland's HERO Parade, the Sydney Gay and Lesbian Mardi Gras Parade, and Pride Scotland. Specifically, in Edinburgh I participated and performed in a women's drumming/marching group in Pride Scotland on 23 June 2001. I draw on this experience and also two focus groups – one with 12 participants, the other with three – and an individual interview held two weeks after Pride Scotland. The data collected on male marching groups comes from Auckland's HERO Parade where I conducted participant observations at three Marching Boys' training sessions, three parades (Auckland and Sydney, 1996, 1997 and 1998) and conducted four interviews with marching boys. Video transcripts and gay newspapers are also analysed in order to argue that all-male and all-female marching groups employ different camp strategies depending on the spatial politics of the parade.

Marching girls

Marching groups are well known not only in Auckland, Aotearoa/New Zealand and Sydney, Australia, but also in northern hemisphere cities such as San Francisco, New York, Toronto, London, and Manchester. Judging by the response of audiences watching the parades from the street-side, marching groups are usually the most popular entry in parades. Marching groups perform a synchronised choreographed march/dance routine to music. Their routines are a parody of 'real' female marching teams that are popular in Aotearoa/New Zealand and Australia (see http://www.marchingnz.org.nz).

In June 2001 I was part of an all women's drumming band that performed for Pride Scotland in Edinburgh. Unlike other marching groups, we played our own music as we carried drums along the parade route. This drumming band formed in January 1996 and performs at various festivals and gigs such as the International Women's Day March in Edinburgh, the famous Edinburgh Torchlight Procession, Hogmanay, Manchester Mardi Gras, Murrayfield Rugby Stadium for international matches, and of course, Pride Scotland. The drumming band carries and plays heavyweight surdos, high-pitched repeniques, snare-drums, tambourines, shakers and bells, to create a samba-type swing sound as they march for Pride. It is this group that I now focus on, before turning to an examination of marching boys.

First, I give context to Pride Scotland and discuss Edinburgh as a major tourist city. Second, I argue that Pride in Edinburgh may be understood as an 'accidental tourist event' that disrupts and challenges the notion of prepared tourist scripts. Finally, I examine how queer women use camp strategies to negotiate hostile and/or indifferent reactions to their bodily performances. I aim to explore how sexualised and gendered power relations (re)form tourist spaces by examining moments where bodily boundaries of sexual difference are established, broken and enforced.

Most of the 20 women drummers that performed in Pride Scotland self-identify as lesbian, some as bisexual, and a couple as heterosexual. Hence my use of the term 'queer', as this diverse group of women live different sexualised lives, but all are critical of gendered and sexualised power relations. Performing in Pride enables an articulation of both queer celebration and protest against normative and oppressive forms of sexuality. On the day of Pride Scotland we met at the assembly point of the parade – the back streets of the Old Town. Approximately 2,000–3,000 other people were gathering and preparing to march for Pride. A decorated bus and truck were the only motorised parade entries as most people were walking. We were a small but colourful and loud Pride celebration.

Our bodies and our drums were dressed in various shades of purple, black and silver, and adorned with glitter. Some of us wore wigs, some of us exposed flesh with sparkly bras and shorts, and others had Xena-like outfits. Some 'butch' lesbians came in drag and wore frocks and feather boas. We looked and felt very camp. Shelia stated that Pride was a special drumming event because:

Shelia: Margaret is more excited usually (laughs).
Moira: It's the only gig with champagne before it.
Lynda: That's important.
Carolyn: It's about celebration.
Margaret: It's a very personal celebration for some. We get dressed up especially for it. We decorate our drums, we decorate our selves. We get more into it, visually.

(First Edinburgh Focus Group, 2 July 2001)

Figure 3.1 Edinburgh Drumming Girls (source: Author's own collection).

Pride parades exaggerate the processes by which bodies and places become gendered and sexualised. At gay pride parades, audiences expect to *see* bodies that defy normative assumptions of gendered and sexualised bodies, while, and at the same time, they may attempt to construct bodies as either masculine or feminine. Gay pride marching teams may be considered as a camp response to the polarisation of appearance and reality, of stereotypes and lived experiences and thereby derive their humour from these opposites. Sontag (1966: 276) writes that camp is 'a sensibility that, among other things, converts the serious into the frivolous' and 'one that is alive to a double sense, in which things can be taken ... between the thing as meaning something, anything, and the thing as pure artifice' (Sontag 1966: 281). While Sontag emphasises the frivolous side of camp, it also has a serious side. Camp may be a means of defiance: a kind of refusal to be overwhelmed by unfavourable odds. It is also a style whereby bodies perform multiple identities, play various parts, and assume a variety of roles, both for fun as well as out of political need.

The need to be camp may come from cities that are inscribed with intense heterosexual rigidity. Sally Munt (1995: 115) notes: 'As I became a victim to, rather than a perpetrator of, the gaze, my fantasies of lesbian mobility/eroticism return to haunt me'. Rewriting herself as a (lesbian) flâneur Munt (1995) challenges the disembodied and masculinist construction of flâneur which is evident in postmodern accounts of 'gazing tourists' (see also Jokinen and Veijola 1997: 23–51, who offer subversive embodied constructions of the flâneur). Munt (1995: 123) argues that in the urban landscape:

> even the protected zones are folding, and yet there are pockets of resistance which pierce the city's metaphoric paralysis with parody: Gay Pride is one such representation, fifty thousand homosexuals parading through the city streets, of every type, presenting the Other of heterosexuality, from Gay Bankers to the Gay Men's Chorus singing 'It's Raining Men', a carnival image of space being permeated by its antithesis.

The drumming group's Pride Scotland performance might be understood using theoretical notions of camp. 'Camp contains an explicit commentary on feats of survival in a world dominated by the taste, interest and definitions of others' (Sontag 1983: 144). Camp performance – through the exaggeration of masculinity and femininity at gay pride parades – has the potential to undermine that which is taken for granted and assumed as natural. Butler (1990, 1993) argues that there is no fixed, real or original gender or sexuality – but that the meaning and interpretation of gender is open to change. 'The excessive performance of masculinity and femininity within homosexual frames exposes not only the fabricated nature of heterosexuality but also its claim to originality and naturalness' (Bell *et al.* 1994: 33).

In our frocks, sparkly outfits and Xena-like iconic costumes, we waited for Pride Scotland to start. Our excitement and nervousness was palpable. We knew that we were near the front of the parade which meant we were first to face people – tourists and locals – on Edinburgh's main streets. Edinburgh is a city with a famous international tourist reputation due to vast amounts of city promotion based on cultural events and festivals (see www.edinburghfestivals.com). Pride Scotland, however, remains marginal to city tourism promotion. This may be due to a construction of Edinburgh as a conservative city when it comes to matters of sexuality. Larry Knopp (1998) uses newspaper articles to discuss discourses of conservatism around an alleged Edinburgh sex scandal involving the judiciary, homosexuality, and various gay cruising sites of Edinburgh. Liz Bondi (1998), in her work on gentrification in Leith, outlines the ways in which local popular press construct women as either promiscuous sex workers or non-sexual mothers (this is discussed further in Chapter 6).

The discursive construction of Pride Scotland in local newspapers is not favourable. One of my participants said:

> *Stella:* The impression that I get is that over the years the media coverage [of Pride] is virtually non-existent. I mean, I certainly, in the Sunday paper, I've only ever seen one article, which was

offensive, in *Scotland on Sunday* which I actually wrote a letter in about because that was the only coverage on the Pride march, and it was all just so homophobic, and I just wrote and said 'what is this all about, what're you talking about?' you know, and they printed my letter actually, which was great.

(Individual interview, 9 July 2001)

An understanding of Edinburgh as conservative has material effects, as another participant noted:

Moira: You know how I was dressed up in a wig and everything? And I was goin' out to meet ya, well when I went out to meet you, I felt *soooo* exposed and such a minority. I never felt like that in Glasgow walkin' in the streets like that, going to Pride ... There's much more of a celebration feeling or warmth, a warmth towards it. In Edinburgh I felt much more antagonist energy.

(First Edinburgh Focus Group, 2 July 2001, emphasis in original)

A combination of city conservatism and lack of Pride promotion has lead to Pride being understood as what I term 'an accidental tourist' event.

In an article entitled 'Performing tourism, staging tourism', Tim Edensor (2001) states that tourist activities are informed by the anticipation that tourists will carry out specific roles and mobilise particular dispositions. Being a tourist at a street carnival, for example, requires a repetition of cultural norms. 'Performative norms need to be continually enacted to retain their power and prescriptive conventions and values that inhere in them are rarely disrupted if they are performed unreflexively' (Edensor 2001: 62). Performative norms for tourists at Pride were disrupted. All went well as we drummed and walked up and along the Mound then down beside the National Gallery of Scotland. When we turned into Princes Street – the main street of Edinburgh – we faced people lined up behind barriers with confused or 'blank' looks on their faces (see Figure 3.2).

The leader of our drumming group, Moira, explains:

Moira: I was terrified when we turned around to go down the hill to go onto Princes Street. I was, and of course we were in a group, we were in the front, but I found it really like *oh my god*, what kind of reaction are we going to get here from these people?

Lynda: I can just remember when we turned into Princes Street and those faces were just looking at us.

Susan: I was actually amazed that there was as many people there were given the amount of publicity there had been. (Shelia: That's true.) So obviously the publicity was all kind of internal, in a sense. So the public interface was, to me, almost accidental.

38 Bodies

Figure 3.2 Pride Scotland on Princes Street, Edinburgh (source: Author's own collection).

Carolyn: Oh yeah it was.
Susan: It seemed accidental.
Barbara: Yeah I agree with that.
Sally: I asked lots of people if they were coming to it and they went 'Oh, I didn't even know about it'. So if you were sort of interested, or you were a lesbian but didn't go to the gay venues, you wouldn't know about it. Nobody knew about it at all.
(First Edinburgh Focus Group, 2 July 2001, emphasis in original)

People on Princes Street, therefore, found themselves as spectators to an event that most, it seems, did not have prior expectations of. Gay pride parades have a history of being exclusive affairs, designed to reinforce communal solidarity between gays, lesbians, bisexuals and transsexuals. In recent years, however, the commodification of Pride has meant that these events – such as Sydney Mardi Gras, Manchester's Mardi Gras, London Pride – have become staged carnival-type performances for large numbers of queer and straight tourists. On Edinburgh's Princes Street onlookers either failed to understand the resonance of Pride or were disparaging of the need to celebrate queer sexualities. As members of the drumming group we were overwhelmed by what appeared to be a city inscribed by intense heterosexual rigidity.

> Kate: I thought it was really strange going down Princes Street because there were so many people there that were so *passive* about what was happening. I don't know if it was because they were tourists and they were just sort of 'oh a spectacle', like they didn't even seem to register necessarily what was going on.... And that, and it was really quite obvious the difference between those people and the people who like were on the tops of buses and other people who had an opinion about it in one way or another and expressed that. So the division between how people perceived it I thought was odd.
> (Second Edinburgh Focus Group, 9 July 2001, emphasis in original)

The resultant ambiguous performance of tourists when faced with our drumming group and other Pride performances on Princes Street threatens, I think, a sense of well-being that is one of the main aims of tourism – to have fun, to let go, to party. Pride Scotland, to these tourists, engenders self-doubt which is not conducive to having a good tourist time. Other spectators displaying disgust at queer women's bodies reinforce hegemonic sexualised power relations. Kate begins:

> Kate: So we were waving at this woman. Sometimes people would ignore you and sometimes they wouldn't, right? But there were some younger people on a bus and they were being really rude ...
> Shonagh: Yeah. They were giving that sign basically. Yeah. It just looked like younger people were just basically saying 'fuck off'.
> Kate: And I think that on that same bus was that older woman?
> Shonagh: Yeah ... she was blowing us a kiss.... And then when you blew a kiss on ... and I copied you ... and she went '*oooo*' like that.
> Erin: She had a really disgusted look on her face.
> Shonagh: Yeah, I thought she was going to *retch*!
> (Second Edinburgh Focus Group, 9 July 2001, emphasis in original)

Munt (1995) shows that positions occupied by subjects are never completely fixed. This tension between violence and freedom from oppression is also the topic of Probyn's (1995) response to Bell *et al.*'s (1994) account of 'lipstick lesbians' in heterosexual space. Probyn (1995: 81) states that:

While one could argue that the sight of two women kissing cannot escape the strictures of heterosexual porn codes, we might also think about, include in our theorising, the fact that making out in a straight place can be a turn-on.

I believe camp to be integral to the creation of this tourist space. Queered tourism geographies are produced by shared humour, vocabulary, the playing of music, the marching together in unison, blowing kisses and embodiment of camp. Our drumming group used humour to dispel hostile and/or ambivalent responses. Miranda discusses the reaction by her work colleagues and boss after a short clip of Pride was televised.

Miranda: Three people told me that I was on tele. Barry is whatever two, three steps up from me, and three people said that 'Barry said he saw you on the television'. Okay, whatever. And *he* actually came up in a meeting and said '*I saw you on the television. I saw you. You were wearing some sort of outfit*'. I was wearing a purple t-shirt and a baseball cap, nothing wild! It's a good thing I wasn't wearing the bra top! (laughter).

(First Edinburgh Focus Group, 2 July 2001, emphasis in original)

Being able to joke about people's reactions, indeed about work colleagues' reactions is one way in which heteronormative responses to queer parading bodies are subverted. Miranda's queer body had become the object of her work colleagues and boss's gaze. Their response was to read her body as exhibiting excess – as many of us were – which she thought was humorous because she felt her clothing was fairly conventional. Other lesbians talked about wearing sunglasses to maintain anonymity in spaces where they might be recognised.

Stella: You know it's like the first time I went [to Pride London] I was so panicked that I was going to be on the tele and 'cause I told my mum I was going shopping in London, you know, and I mean I obviously still never tell her where I'm going or whatever.
Lynda: Do you not?
Stella: No and when I went in Pride Scotland in Glasgow I had to go with sunglasses on because we passed my dad's office, which is you know like pretty grim really.

(Individual interview, 9 July 2001)

Stella's example highlights an ironic strategy of (in)visibility. While she is out to her parents, the risk of being recognised by her parents, or people who know her family, may result in family disapproval. Her desire to be involved in Pride events, however, brings into tension the dynamics of pride and shame. As Munt (1998: 97) has noted: 'pride is dependent on shame; pride is predicated on the – sometimes conscious – denial of its own ostracised Other'. In more extended terms, attention to pride *and* shame may allow new insights into the normalising categorisations of bodies and spaces. Stella expresses more about the relationship with her parents.

> *Stella:* Oh I mean it's not anything specific purely specific to Pride. It's just like I came out seven and a half years ago you know. There's just no going anywhere with that really with my Mum and Pop. So we don't talk about and I don't, I hate talking about it and my mum certainly isn't in to it and we've had an argument once about, she said 'why do you feel the need to a) tell everyone and b) walk down the street'. Her exact words were that 'you've got the exact same rights as everyone else you know don't shove it down their throats sort of thing'. And that's a few years ago and I don't know whether she's changed her stance on that or not. I mean my mum says that if I get beaten up it's my fault for telling people about it, rather than their fault . . .
>
> (Individual interview, 9 July 2001)

Stella, despite this fraught relationship with her parents is still resolute about her role in Pride when it is held on home ground in Glasgow.

> *Stella:* It is really important for local, for every one to have their own kind of celebration or whatever. Um I guess it makes me a bit more vulnerable as well because you know I'm a bit [less] anonymous, because I know loads of people who could walk through the streets and not meet anyone that they know, but definitely in Glasgow I am much more exposed, you know.
>
> (Individual interview, 9 July 2001)

In conclusion to this section on marching girls, the performative tourist space of Pride Scotland highlights the difference that space makes to camp performances. As a result of the lack of Pride promotion by Edinburgh city and Pride officials, tourists are 'caught by surprise' without a tourist script and Pride performers are suspicious in what I call an 'accidental tourist event'. Experiences of queer women who participate in Pride Scotland draw on camp sensibilities in order to challenge heteronormative understandings of bodies and spaces. Butler (1990, 1993) recognises that gendered and sexualised transgressions are regularly punished and power relations that inscribe bodies and tourist spaces can be hurtful and shaming. Diverse strategies are employed by queer women in Pride, such as using humour, exaggerated play of femininities and masculinities, and being present/absent by hiding one's personal identity.

In the remainder of the chapter I turn to male marching groups of Auckland, New Zealand and Sydney, Australia to highlight the ways in which Pride events queer tourism encounters. These parades might be understood as 'directed performances' (Edensor 2001: 73), conducted on a type of tourist stage that makes parades highly commodified by cities, pride officials and corporations. This type of tourist stage means that the enactment of camp produces different embodied performances.

Marching boys

The male marching groups that I have seen perform tend to have between 40 to 70 members. Their routines usually have moments when they are stationary and turn directly to face the crowd. They also march forward so as to advance along the parade route with the other parade entries. Routines tend to be performed to music that is significant to gay male communities. For instance, overtly 'gay music', such as 'YMCA' from the group The Village People is a popular tune to march to.

Male marching groups often wear theme related 'uniforms', for example, the Locker Room Boys at the 1995 Sydney Mardi Gras wore gym towels because they wanted to affirm and celebrate 'gym' culture. In 1996 the HERO Marching Boys wore a lifesaving outfit which incorporated swimming shorts, goggles and caps. In 1997 the HERO Marching Boys wore sparkly shorts and narrow wrap around sunglasses as a way of adopting a futuristic 'outer space' theme. Marching boys at the 1997 Sydney Mardi Gras Parade wore pink shorts with a heart shape cut-out in the buttocks. Many of the uniforms worn by the marching boys are 'brief' and expose lots of skin and muscle.

The following excerpt is from the ABC televised production of the 1995 Sydney Mardi Gras Parade. ABC television presenters Angela Catterns, David Marr and gay comedian Julian Clary, comment on the parade as it passes through the streets. Another ABC television presenter, Elle McFeast, mingles and talks with the paraders on the street:

Angela Catterns: Here are the Locker Room Boys. They've liberated the locker room. I'm sure Elle [McFeast] is going to find out what is under their towels.
Julian Clary: Explain this to me. I don't understand being a foreigner. What is a locker room?
David Marr: It's a changing room attached to a – by the looks of them – a gymnasium. They're wearing their towels.
Julian Clary: I think it's positively bizarre.
Angela Catterns: The idea is to take the locker room out of the gym and the sauna domain, Julian. They want to take it out and march it proudly along the street and have fun.

Julian Clary: But women go to the sauna, women go to the gym.
David Marr: They don't go into the men's changing room – even in this country . . .
Julian Clary: Ohh, they have taken their towels off . . . I shall have to cover my eyes in a minute.
(*Sydney Mardi Gras Parade 1995*, ABC Video)

Becoming a Sydney Mardi Gras or HERO marching boy usually involves engaging in specific exercises, usually carried out at a gymnasium, in order to become muscular. Having a muscular body – in western traditions – has tended to be a masculine pursuit. Body building for men can be seen as the fulfilment of a hegemonic notion of masculinity and/or virility. Male body building can be read as an attempt to render the whole body into the phallus, 'creating the male body as hard, impenetrable, pure muscle' (Grosz 1994: 224). Strength, stamina, control and virility are all attributes associated with muscular male bodies and appear to be sought by the Locker Room Boys' bodies. The connection between muscularity and maleness often becomes naturalised and essentialised.

It is possible to read the bodies of the Locker Room Boys and the HERO Marching Boys as a camped up enactment of masculinity. A photograph of the HERO Marching Boys appeared in the *New Zealand Herald* (1997: 20) under the caption 'Marcho Men'. The caption 'Marcho' is a play on both macho and marching (see Figure 3.3). The emphasis on the masculine (macho) pertains to the corporeal specificity of their bodies. The photograph

Figure 3.3 The HERO Marching Boys: 'Marcho Men' (source: *New Zealand Herald*, 24 February 1997: A20).

of the HERO Marching Boys depicts bodies that are hard, muscular, and masculine. The bodies are broad shouldered and small waisted. This creates a V shaped torso that is a traditional marker of masculinity. The lack of clothing (usually the HERO Marching Boys wear small tight shorts or some variation on this theme) exaggerates the crotch and accentuates stomach, thigh and buttock muscles. These bodies are produced through the 'calculated tearing and rebuilding of selected muscle groups' (Grosz 1994: 143) and through a marching routine that 'pumps' their muscles. The pump is the result of high intensity training and muscle stimulation. Muscles become engorged with blood and short of oxygen which causes the skin to stretch tight over the muscles. The pump normally results in muscle growth because more blood flushes through the muscles.

An indicator of the HERO Marching Boys' popularity as entertainers can be found in a report in the *Man To Man: New Zealand's National Gay Community Newspaper* entitled 'Warriors court marching boys' (Bennie 1995: 1). The 'Warriors' – who are the Aotearoa/New Zealand national representative male rugby league team – approached the HERO Marching Boys looking for a complement to the usual (and 'traditional') marching girls which perform to the audience before the rugby league game begins. The HERO Marching Boys' spokesperson Robert Chung said, 'I was very flattered when the call came through' (cited in Bennie 1995: 1). Bennie reports that: 'The [HERO] Marching Boys were one of the hottest acts of this year's HERO Parade down Auckland's Queen Street. Fans, many of them young heterosexual women, showed their appreciation with enraptured screams' (Bennie 1995: 1).

It seems as though the HERO Marching Boys were approached with an offer to perform because 'young heterosexual women' find their performance enrapturing. This is an example of the way the marching boys' bodies can be read as hypermasculine or 'all-man'. The HERO Marching Boys, however, turned the offer down. Chung stated: 'On reflection we felt that the [HERO] Marching Boys worked well in the friendly context of the HERO Parade, but we were unsure how we would be received by a rugby league crowd ... some of the boys felt it just wasn't the right place' (cited in Bennie 1995: 1). New Zealand rugby league fans at the Auckland stadium, therefore, were not provided with pre-match entertainment put on by the Marching Boys at the rugby league stadium.

Professional male rugby league players and the Marching Boys have several corporeal similarities. Rugby league players are usually very muscular and have little body fat. Their muscles are visible. Their uniforms are tight and reveal flesh. The HERO Marching Boys' bodies, as I have been arguing, are also muscular and they work their bodies to make their muscles 'stand out'. Both body 'types' could be regarded as hypermasculine, or 'all-man', as they are hard – not leaky or soft – bodies. The rugby league players and the Marching Boys are both popular 'body' spectacles, and hence both are inscribed by the heterosexual gaze.

The hard, pumped, muscular bodies of the HERO Marching Boys attract a great deal of attention from the parade crowd. The marching boys move in unison. This gives the appearance of being one large masculine and muscular body. Attributes of strength, stamina and control become inscribed on the bodies of the HERO Marching Boys during the length of the parade. The body discipline needed for being part of marching groups is intimately linked to power. Biopower is a useful concept for understanding:

> the body as a machine: its disciplining, the optimization of its capacities, the extortion of its forces, the parallel increase of its usefulness and its docility ... all of this was ensured by the procedures of power that characterised the disciplines: an anatomo-politics of the human body.
>
> (Foucault 1976: 139)

This type of performance and profile may help establish a popular profile and glamorous status while at the same time the HERO Marching Boys' routine could be interpreted as 'hybrid' activity, part cheerleading and part 'real' marching girl style. Their display, movements and dance can be read as highly 'feminine' and reinforce a camp sensibility. Dyer (2002: 49) claims camp as a survival technique: 'It's being so camp that keeps us going' while Bell *et al.* (1994: 36) argue that this is manifested in 'greater care and attention invested in dress, oppositional or otherwise' (see also Jackson 1991, 1994).

Dwayne, one of the HERO Marching Boys with whom I talked, said that he was attracted by the 'international' status and prestige associated with being one of the HERO Marching Boys. Dwayne explained his involvement in the HERO Parade (please note: (//) indicates an overlap in talk):

Dwayne:	Ah, I think a) they look excellent, um when they do their little routine thing. And um, b) it seems like, for example in Sydney, it's such a, a honour, you know (//).
Lynda:	(//) A privilege really?
Dwayne:	Yeah.
Lynda:	It seems to be a very high profile kind of um, thing to be in the parade?
Dwayne:	It is. Yeah it is, I agree.
Lynda:	Yeah, did that attract you to the job?
Dwayne:	It certainly did, all the glamour.
	(Individual interview, 13 February 1996)

High profile and glamour suggest that the HERO Marching Boys are extremely important to the success of the parade. Dwayne reiterates this position.

46 *Bodies*

Lynda: Why did you want to do it? What got you motivated to be a Marching Boy?

Dwayne: Um, because I'm sick of being a spectator (oh, right), and it looked like fun ... So um and I've seen a few of the parades and I go to Mardi Gras and it's, it's just fun.

Lynda: Yeah. High energy?

Dwayne: Yeah. No political statements.

(Interview, 13 February 1996)

Dwayne just wants to have fun. He suggests that gender/sex roles are a thing of the past:

Lynda: But it [marching] is something that usually a bunch of girls do in dresses?

Dwayne: No I've never really thought that. Because gone are the – what girls do, and what boys do – days.

Dwayne insists that the rigidity of gender roles no longer exist. Is it possible for the marching boys to simultaneously occupy masculine, feminine bodies? Do marching boys embody a new type of masculinity which opens up opportunities for resistance to hegemonic subjectivities? Tourists may reinforce the dominant position of marching boys in the parades and therefore help to normalise and valorise them. This, in turn, may help to (re)position the marching boys as camped up hypermasculine. National television coverage of the 1997 HERO Parade reflects the reaction of spectators towards the HERO Marching Boys:

Anita McNaught: They are a firm favourite with the crowds on the HERO Parade. They were here last year, they were here the year before that. They are now 50 strong and that's a record.

(Television 3 Network Services Limited, 1997)

Furthermore, the HERO Marching Boys are the only team to officially represent Aotearoa/New Zealand in the Sydney Mardi Gras (see *express* 21 February, 1998). This is another indication of the dominance and prestige that the marching boys embody.

When I asked Malcolm and Brad why the HERO Marching Boys are so popular, they responded that:

Malcolm: Cos they're all spunks.

Brad: Cos they're all like sex symbols. They're half naked and most of them are pretty good looking.

Lynda: Yeah. Good bodies?

Brad: Yeah and pretty skimpy shorts.

(Group interview, 13 February 1996)

Despite their attempts at displaying hypermasculine, or 'all-man' bodies, and maintaining a position of dominance within the parade, the Marching Boys can also be conceived of as feminine.

Marching boys devote a substantial amount of time and energy to making their bodies 'right' for gay pride parades. This attention to the appearance and performance of their bodies could be understood as a feminising process.

In the *Sydney Morning Herald*'s (1998) feature article on the Sydney Mardi Gras several body tips were disclosed for getting ready for Mardi Gras. Under the heading 'Prepping Up' was the following advice to recognise the importance of the event:

> Somewhere between obsession and slovenliness lies a healthy desire to look one's best for an important occasion. And for gay men and lesbians in this town, occasions don't come any more important than Mardi Gras night.
>
> (*Sydney Morning Herald*, 3 March 1998)

Working out at the fitness centre, swimming, running and attention to diet were all recommended to create a body to be proud of. The article also covered procedures for tanning, waxing, tattooing and piercing. Furthermore, shopping for the parade outfit to complement the worked body was also listed as a high priority in the article.

Shops specialising in 'skimpy shimmy shorts' did a brisk business in Sydney prior to the Mardi Gras. The bodies being prepped for the parades are reconstituted through traditional feminine practices. Several beauty companies advertised their services prior to the HERO Parade. One company, Adeva Esthetique, specifically targeted gay males in time for the HERO celebrations. Their advertisement read:

> Get off my back this HERO! If you've got hair in places you could do without, why not lose it completely. We offer a safe, hygienic and private waxing/clipping service that'll get the hair off your back and maybe someone new on your tail. Back waxing, Aromatherapy, Facials, Massage. For bookings just give Bradley a call on ...
>
> (*express* 1996: 13)

To fulfil the requirements of becoming a member of the hard-bodied marching boys, they also had to become feminised through the effects of training, tanning, and waxing. The body becomes marked with the use of makeup, hair products, clothing, posture, diet, weight training, and aerobics. Several of the HERO Marching Boys were aerobic instructors.

The HERO Marching Boys team trained for several months prior to the parade. When I observed training sessions in Auckland, they had their own female choreographer who regularly trains female ('real') marching girls' teams. Through habitual (feminine) patterns of movement and exercise, the marching boys sculptured their bodies for the parade. As well as the hard work that went into the training, the sessions were extremely enjoyable, involving plenty of jokes about waxing bodies and getting hair done and so on.

Their training and marching routine is designed to accentuate and build their bodies. The performance is high energy and necessarily predicated on exposed flesh. Another reading of the marching boys' bodies is that their small waists, curvaceous pectoral muscles, hairless and oiled flesh are all corporeal indicators of femininity. Furthermore, the process of obtaining a carefully crafted body is a feminising and disciplining activity. The Sydney Mardi Gras Locker Room Boys are willing to experience pain in order to obtain the ideal body. In the 1995 Mardi Gras Parade, television presenter, Elle McFeast, interviews one of the Locker Room Boys:

> *Elle McFeast:* Tell me – how important are gyms to the gay culture?
> *Locker Room Boy:* We are the gay culture, we are the essence of the gay culture.
> *Elle McFeast:* A lot of waxing has gone into preparations?
> *Locker Room Boy:* Yes, we have gone to a lot of trouble, a lot of effort, a lot of pain, to make people happy.
> (*Sydney Mardi Gras Parade 1995*, ABC Video)

The evocation of femininity (through training, routines, prepping and so on) enables the bodies to be built to a 'parading' state of hypermasculinity. As a result, these marching boys' bodies also confuse traditional or 'natural' corporeal indicators of masculinity.

There are moments of contestatory politics and other paraders have reacted to the status of the official marching boys. One such reaction was a parade entry called the *express* Check-Out Chicks. Robert, a member of the *express* Check-Out Chicks, described their intentions:

> We are twelve drag queens or so, and it's almost a parody on the [HERO] Marching Boys and it's a full parody of like, like, normality. We're taking like Foodtown [supermarket] check-out trolleys and marching.
> (In-depth interview, 7 February 1996)

The performance of the HERO Marching Boys is subject to another performance. Like the HERO Marching Boys, the Check-Out Chicks use whistles in their routine, but rather than 'strip down', like the HERO Marching Boys, the *express* Check-Out Chicks 'dress up' in full drag. They explicitly hyper-feminise themselves in an attempt to send up the hypermasculinism of the HERO Marching Boys. The supermarket trolleys also feminise their performance and enhance the parody. The name 'Express' was a play on their sponsor, a newspaper called *express: new zealand's newspaper of gay expression*. 'Express' is also a reference to the express, or fast-track check-out lanes at the Foodtown supermarkets. Shopping at supermarkets can also be understood as a feminised activity, or what Robert calls 'normal' activity. There are two things worth noting here. The first is that spectators are not expecting to see 'normal' or 'everyday' activities such as shopping in a gay pride parade which may upset their notions of the Other. The second point that is worth noting is that this is a parodic repetition of a parodic repetition – in other words, this performance is a parody of a parody, in that the regulatory forces that shape marching groups, are camped up by the HERO Marching Boys' bodies, and in turn, restyled by the *express* Check-Out Chicks. One outcome of this might be that any sense of an original gender performance is thoroughly misplaced and reimagined. 'Understood at least partially as a political act, camp drag produces an effect that will critically comment on or at least inflect the gender system of a particular historical/cultural moment in unpredictable, but readable, ways' (Piggford 1999: 288). Marching boys play with their respective cultures as well as with gay pride audiences. In this way, gendered and sexualised bodies are produced in relation to each other, as well as in relationship to the queered spaces of gay pride parades.

The transgressive possibilities of gender performance may lie somewhere in between the Pride performance and the tourists' intentions. When I spoke to the HERO Marching Boys, it became evident that their routine 'worked' because they played with, and seemed excited about, heterosexuals watching and constructing them as Other. The HERO Marching Boys looked back at the spectators. They also directed sexually suggestive movements and expressions towards the spectators. Several HERO Marching Boys lead the routine and 'turned up the crowd':

Malcolm: We're supposed to be the young, young dancing boys.
Brad: Sort of turns everybody on. We're supposed to turn up the crowd.

(Group interview, 13 February 1996)

The performance of the HERO Marching Boys, in fact, relied on heterosexual audiences. Malcolm and Brad turned up the crowd, and each other, by dancing together and by breaking through the road barriers and dancing with male and female spectators. In these body touching moments, spectators and marching boys disrupt, and hold in tension male/female and heterosexual/homosexual binaries. The spectators are no longer 'safe' behind the road barriers from queer bodies. Rather, they become part of the spectacle when they are drawn into the performance with the HERO Marching Boys. The actions of the HERO Marching Boys, for a queer sexed space moment, unhinge heteronormativity. Their exaggerated and camp performances invite tourists to also perform and play. Tourist space is reproduced by tourists, who perform a dramatised response to Malcom and Brad. Gay pride tourists are shaped by an orientation towards the kinds of experiences on offer.

Gay pride parades, therefore, become encounters of freedom *and* oppression. Parading bodies cannot undo the historicity of the ways in which heterosexism produces a place for the production of the Other. The fact that queer bodies materialise their same sex desire, however, can go some way towards rearticulating that site/sight. It could be argued that the performance of queer desire for heterosexuals is a turn-on. Lustful same sex displays in gay pride parades can be thought of as 'a type of articulation of desire that bends and queers' (Probyn 1995: 81) a (heteronormative) gaze, allowing for a momentarily sexed queer space.

The conflict between paraders and the ensuing parody upon parody of subjectivities produces a complex tourist space. Who desires this spectacle? For the spectators the parading bodies might open up the tourist space as 'not only multiple, composite, heterogeneous, indeterminate and plural' but also as 'space with dimensions of which cannot be completely described, defined, discoursed' (Rose 1997: 188). Rose (1997) asserts that particular imagined spatialities are constitutive of subjectivities. Identities are in part constituted by the space through which they imagine themselves.

The television commentary for the 1997 HERO Parade was provided by a television personality – Anita McNaught – and the lesbian performance duo, Lynda and Jools Topp in their masculine personae of 'Ken' (Lynda) and 'Ken' (Jools). 'The Kens co-commentated, heavily working the irony of having a couple of lesbians playing the sort of guys who, in real life, would consider a suitable finale to the HERO Parade to be a well-aimed lightning bolt from God' (Wichtel 1997: 64).

Ken (Jools Topp): Well, it looks like they've got a great routine worked out Anita. Do they spend a lot of time on it?

Anita McNaught: Apparently so – they rehearse to within an inch of their lives, those boys.

Ken (Jools Topp): Oh – it's great to see that, isn't it Ken?

Ken (Lynda Topp): Oh it is Ken. And good to see, ah, blokes out there actually cheer leading for a change Ken and, ah, by God they do it well don't they?

Anita McNaught: They put in about 25 hours practising their routines over the last, over the last month or so.

The results of this disciplining are displayed through their bodies:

Ken (Lynda Topp): It looks like there's been a bit of working out ... Look at the muscles on some of those Anita.

Anita McNaught: There are some fine, fine looking men there.
Ken (Lynda Topp): Yeah, very athletic.
(Television 3 Network Services Limited, 1997)

The training of the HERO Marching Boys creates bodies which can be read as both feminine and masculine. The evocation of femininity (through training, routines, prepping and so on) enables the bodies to be built to a 'parading' state of hypermasculinity. As a result, these marching boys' bodies also confuse traditional, or 'natural' corporeal indicators of masculinity.

While the HERO Marching Boys are a well-established parade favourite, the HERO Marching Girls, however, are less well known. When they do perform they attempt to undo the regulations of conventional and traditional (heterosexual, patriarchal) femininity. Activities such as dieting, hair removal, dressing-up, make-up, and so forth, are not prerequisites for becoming a member of the HERO Marching Girls. The Marching Girls could be understood as lesbian feminists who reject patriarchal notions of femininity because it is seen as a symbol of women's oppression.

> Femininity and its trappings thus came to represent to some feminists women's oppression. Feminism sought to separate femininity from the woman or gender identity from the sexed body and return it to the 'original, natural' woman underneath all that make-up.
> (Bell *et al.* 1994: 41–42)

With this stance there is an implicit rejection of those lesbians who might think body size is important, and who might engage in dieting, hair removal, make-up, maintaining low body fat, the body aesthetics of gym training and so on.

> The success of dress as a transgression of heterosexuality/patriarchy and heteropatriarchal space is reflected in the fact that dungerees (overalls) and Doc Marten boots became for the popular press synonymous with feminism, feminism with lesbianism – a success which could be measured in the popular press backlash against these forms of dress (Wilson 1988; A. Young 1990) and the abusive and hostile reactions of men in public space to women dressing in this way.
> (Bell *et al.* 1994: 41)

The HERO Marching Girls gained sponsorship from 'Doctor Martens' and subsequently wore Doc Marten boots in their marching routine at the HERO Parade. They wore full-bodied skin coloured leotards and tights with skirts, and did not 'show' any flesh apart from bare arms.

Differences between paraders may not only be based on gender, but also on age. Probyn (1993: 10) discusses some of the politics of group identity surrounding her experience at Gay Pride in Montreal:

As we proceeded along our route, it became obvious that many of these marchers were from another generation – the ten years separating me from these proudly 'out' dykes in their early twenties meant that the 'we' somehow died in my throat along with the chant, 'We're here, we're queer, we're fabuuulous'.

Marching bodies on parade at HERO and the Sydney Gay and Lesbian Mardi Gras, demand to be rethought in terms of sexual differences, sexual practices, desire and discourse. Thinking about the enactment and exaggeration of gender and sexuality, or camp performances, is a useful way to reconceptualise these tourist events as non-hierarchical, permeable, and paradoxical.

Conclusion

I have teased out some of the ways in which the bodies of marching groups are gendered and sexualised at Pride Scotland, HERO and Sydney Mardi Gras parades. Social meanings produced in these parades – in which the paraders, who are both shaped by these institutions and are agents of change – produce changes which may both serve hegemonic interests and challenge existing power relations.

These parading bodies are always sites of contradiction and are constantly reconstituted in dominant, subversive and competing discourses. My project here has been to decentre the subjects of the parade by abandoning the belief in an essential subjectivity. Rather I establish a changing and shifting subjectivity. These conflicting meanings become written on bodies.

The construction of marching bodies can be understood as performing multiple types of masculinities and femininities. Cream (1995: 31) asks: 'What is the sexed body?' She argues that this question is rarely asked, nor can this question be easily answered. Bodies are sexed and sexualised by places.

Bodies on parade, such as the Edinburgh marching/drumming girls, HERO Marching Boys and the Locker Room Boys of the Sydney Mardi Gras Parade, demand to be rethought in terms of sexual differences and sexual practices. Political change involves a struggle over meaning; therefore, it is not possible to dismiss gay pride parades as merely servicing or opposing the dominant and oppressive (dualistic) systems of western thought. Consequently, these gendered bodies are always open to challenge and redefinition with shifts in their discursive contexts. Therefore, at any given moment, bodies are open to constant rereading and reinterpretation. Over the last decade there has been a persistent call to examine alternative frameworks and discourses for capturing the performativity, diverse and contradictory ways in which sexed and sexualised bodies are made. My account of parades and bodies may be one way to do this.

I have adopted notions of camp and some of Butler's (1990, 1993) work to highlight the performativity of gendered and sexualised bodies in order to explore the difference that space makes to camp strategies. Experiences of Edinburgh's marching girls suggest a very different tourist space from the marching boy spaces of Auckland and Sydney. As a result of tourists' indifference or hostile reactions, marching girls use camp to make fun of and undermine the power of homophobia. Through the meshing of humour and politics the women's drumming/marching group is able to challenge Edinburgh's conservative street spaces.

The male marching groups enter into street spaces which are already staged for spectacular displays of flesh and fun. These tourist spaces allow and encourage exaggerated performances of gender and sexuality. Parody enables gender play to a point that original meanings are invalid and multiple aspects evolve.

I have teased out some of the ways in which marching bodies are gendered and sexualised at Pride Scotland, HERO and at the Sydney Mardi Gras parades. Bodies on parade, such as the women's drumming group, HERO Marching Boys, *express* Check-Out Chicks, and the Locker Room Boys of the Sydney Mardi Gras Parade, upset fixed ideas of sexual differences and gendered bodies. By reading the camp performance of marching bodies as masculine, feminine, and as *both* masculine and feminine, I hope I have shown bodies to be volatile, mobile and contradictory. The deployment of camp serves to undermine the naturalness and originality of gendered bodies and to reveal gender as performative. The binary man/woman becomes a regulatory fiction when queer bodies camp up notions of femininity and masculinity. Social meanings produced in these parades – in which the paraders, who are both shaped by these institutions and are agents of change – produce changes which may both serve hegemonic interests and challenge existing power relations. A focus on embodied performances at Pride challenges popular held assumptions about gender and sexuality because they parody gender heterosexual identities *and* homosexual identities. Their bodies become readable in a myriad of ways that may create anxieties and trouble. Writing about camp bodies and spaces challenges some of the fleshless and static territory of tourism studies.

4 Street scenes
Tourism with(out) borders

This chapter examines the discursive and material street 'borders' at gay pride parades. I am interested in the ways in which borders – that organise and give shape to tourism experiences of gay pride parades – are discursively unstable. When the borders between paraders and tourists wear thin they threaten to disrupt the social order of tourism space. The securing of borders around bodies (queer and heteronormative) and places (particular streets) are attempts to ensure a controlled tourism experience. These borders come loose, however, despite efforts to secure them. 'Loose borders' both excite and repel various tourists at pride parades. Furthermore, particular bodies and performances that threaten the borders of corporeal acceptability also may threaten the heteronormative space of streets (Johnston 2002).

Some bodies on parade, for example, bodies involved in bondage and sadomasochism (s/m) performances, bodies living with HIV/AIDS, and bodies which contest the gendered corporeal borders, can be partially explained using Iris Young's (1990a) notion of cultural imperialism. These bodies are constructed as 'freaks', 'ugly', 'dirty' and are further Othered by particular watching tourists. Julia Kristeva (1982) also examines the notion of 'abjection'. Kristeva questions the conditions under which the proper, clean, decent, and law-abiding body is demarcated. This theory helps make some sense of tourists' and paraders' reactions. There is a fear that some bodies in the parade are too risqué and therefore can't be trusted in public streets. Contradictions and ironies structure the analysis of the empirical material in this chapter in that spectators are both drawn to, and at the same time, repelled by bodies on parade. I draw on heteronormative tourists' responses to a questionnaire in order to argue that the dominant, unmarked position of the tourist is maintained through discourses of liberalism. There is potential, however, to disrupt this unmarked position and question the binary between self/Other, tourist/host, and straight/gay by examining specific parade entries, such as Gaily Normal, Rainbow Youth and the Gay Auckland Business Association (GABA). Binary oppositions in tourism discourse, such as self/Other, straight/queer and tourist/host are troubled, as these groups perform and parody heteronormativity and hence have potential to destabilise dominant meanings of homosexuality/heterosexuality, host/tourist.

Bodies on display for tourism depend on particular commodifiable experiences. Within a context of gay pride parades, a tension and contradiction is maintained between parading bodies as tourism commodities, and political displays of pride. It is within this contradiction that this chapter is situated. Parading bodies become borderline bodies – bordering on politics, streets, and tourism practices.

This chapter is structured in the following way. It begins by situating parades within discourses of carnival. Taking the starting point that parades are simultaneously forms of tourist entertainment, as well as political protests, I discuss the spatial constitution of symbolic resistance through the lens of the carnivalisation of society. Streets can be queered and, as a result, heteronormativity becomes denaturalised – made obvious – during gay pride parades. Parading bodies in streets can be understood as making queer bodies seem *queerer*, which reasserts Otherness. Second, I draw on empirical data from Aotearoa Auckland HERO parades and Sydney, Australian Gay and Lesbian Mardi Gras parades, in order to develop my argument in suggesting that queer parading bodies can arouse feelings of abjection which play a crucial role in the popularity of the parades. I use the notion of abjection because it is a useful theoretical and political strategy. Parading bodies that are abject trouble borders at the same time as the Cartesian subject tries to redraw them. I focus on the urge to separate self/Other and straight/gay, and the mobility of subject positions that this urge creates. The intention for this chapter, therefore, is to trouble the notion of fixed subjectivities and fixed spaces by examining embodied contradictions.

Spectacular carnival events

While gay pride parades have their own history, they may also be understood as socially contested events whose political significance is inscribed in the landscape. The presentation of marginalised peoples in public places can be seen as subversion and transgression of what is usually termed 'high' and 'low' values. This presentation, however, is more than a site of hierarchical social inversion. It is worth noting that HERO and Sydney Mardi Gras are parades of the 'South' and are very different from comparable events of the 'North', which tend not to be structured around entertainment and difference. This chapter elaborates on the place and political specificities of gay pride parades 'down under'.

Claire Lewis and Steve Pile (1996: 39) state: 'In certain places and in certain times, carnival may be a ritualised resistance, or it may be a contested territory, or it may be a site of hybrid ambivalences, or it may be an opiate to the people'.

Carnival has been attributed with providing an opportunity for self-expression among marginalised groups, as can be seen in the Notting Hill

Carnival in London (Cohen 1980, 1982; Manning 1989), and there have been several studies of the construction of gender and race at carnival events (see Jackson 1988, 1992; Lewis and Pile 1996; Spooner 1996). The idea of carnival as reversal establishes the dominant social order as something which is static and is 'allowed' to be temporarily punctured. Carnival is understood as 'a *licensed* affair in every sense, a permissible rupture of hegemony, a contained popular blow-off as disturbing and relatively ineffectual as a revolutionary work of art' (Eagleton 1981: 148, italics in original).

Carnival, therefore, can be seen as a form of ritualistic value display which redefines the meaning of urban spaces. These displays have often been discussed in terms of urban economic gain versus possible civic disruption, which tend to place city governments in a double bind. On the one hand they are interested in events that make the city attractive to large numbers of people, because money spent at events indirectly feed tax revenue. On the other hand, they perceive such events as threats to the establishment because they are often spatially unstructured and involve large groups in playful activities (Bonnemaison 1990; Hall 1992). The political impact of the Sydney Mardi Gras affects many areas of public and civil functions, from, for example, political consciousness and organisational ability for lesbian and gay communities, to major political centres and the state.

The representation of carnival as the reverse of static, everyday normality has been an important starting point in a focus on the spectacular. Attention needs to be paid, however, to the contradictions of carnival bodies as constituted by and with the spaces in which such events take place. HERO and Mardi Gras parades are also contained within dominant discourses of tourism and entertainment, which tend to create binaries of self/Other, or tourist/host. Furthermore, Sydney Mardi Gras and HERO parades are not just a one night affair. Their meanings are constructed from a month-long festival of events which involve, for example, art exhibitions, theatre, film festivals, health programmes, and tourism.

Redefining urban space

Gay pride parades do not simply (and uncontestedly) inscribe streets as queer, they actively produce queer streets (Bell and Valentine 1995). Parades can be read as deconstructive spatial tactics, a queering of streets. Nancy Duncan (1996: 139) states 'Gay pride parades, public protests, performance art and street theatre as well as overtly homosexual behaviour such as kissing in public' upset unarticulated norms. Munt (1995: 124, emphasis in original), argues that such behaviours produce a 'politics of *dis*location'. Duncan (1996) believes that lesbian and gay practices, if they are made explicit, have the potential to trouble the taken for granted heterosexuality of public places. The street tactics of gay pride parades are also 'crisis points in the normal functioning of "everyday" experiences' (Cresswell 1996, cited in Duncan 1996: 139). Valentine (1996: 152) argues that 'Pride marches also achieve much more than just visibility, they also challenge the production of everyday spaces as heterosexual.'

Discussions of the vitality of pride parades do not usually focus on the role of tourists and tourism industries in such controversial displays in public space. When gay bodies are on parade, however, they are clearly marked as 'different'; their bodies constitute an 'extraordinary' tourist attraction. This is because normative notions of ten result from a binary division between the ordinary – so called 'everyday' – and the extraordinary. The dichotomies of tourists and hosts, or self and Other, however, are not inherent or 'natural' binary divisions. They are produced, for example, when bodies become gendered and sexualised at gay pride parades. Away from the parade, the tourist event, queer bodies may seem 'ordinary', 'everyday', and even 'normal'. During parades, however, the border or binary division becomes visible and accentuated between paraders and spectators.

The creation of such oppositions provides a spectacle of queer bodies to the dominant culture. That the majority of tourists at HERO and at the Sydney Mardi Gras are heterosexual is strongly suggested by a questionnaire survey I conducted among spectators on the night of the 1996 HERO Parade: of the 114 people returning questionnaires, 90 (76 per cent) identified themselves as heterosexual. (Of the remaining responses 18 (of 114) identified as gay male, three identified as lesbian, three identified as bisexual and four did not specify their sexuality.)

The popularity of gay pride parades for heterosexual spectators, particularly the Sydney Gay and Lesbian Mardi Gras Parade, has been noted elsewhere. Bell and Valentine (1995: 26, italics in original) claim:

> It seems that the construction of Pride marches for a straight [tourist] spectator audience is becoming a very important issue for marches in the US and, judging by some footage of Mardi Gras shown on British TV recently, in Australia too (look at who's *watching* the parade).

HERO, according to the HERO Project Director, has always been constructed for the 'straight' tourist spectator:

> Well the parade is basically put on for the straight community when it comes down to it. Like a hundred thousand people there, I don't know, 5000 would be gay? ... Ah, so it's for straights and that's fine. I don't think we should have a problem with that at all. We should encourage it.
>
> (Individual interview, 22 September 1995)

Another indicator of the construction of HERO for the 'straight' or normalised public is that since 1997 the full parade has been presented on prime-time national television. The HERO Parade as public 'product' is now sold to television production companies and can be purchased as a video cassette. The Sydney Mardi Gras is also televised in Australia and marketed in video cassette form. Such products are advertised as tourist souvenirs in Sydney, along with T-shirts, tea towels, key rings and so forth.

To summarise the discussion so far, gay pride parades have the potential to queer the 'everyday' or 'ordinary' streets of cities (also see Johnston 1997, 2001; Brickell, 2000). They also tend to be caught up in discourses that construct queer bodies as Other, and tourists as self. The spectacle of the HERO Parade is constituted through binaries of tourist/host, and straight/gay. Tourism literature suggests that these events can be theorised as the powerful producer of the exotic (Rossel 1988) and a commodifier of cultures (Greenwood 1978) which constructs 'Others'. This contradiction troubles the spaces and bodies at gay pride parades.

Placing borders

HERO and Mardi Gras are intensely structured spatial events. Clearly marked borders between paraders and tourists are maintained at the roadside through the use of road markings, road barriers or barricades, parade officials and police, as well as self-policing by tourists. I am referring to general social control and regulation of communities which is carried out by all of its members. Individuals also engage in self-surveillance, self-control and self-disciplining regimes (see Foucault 1970). This may be one the reasons why the HERO Parade and the Sydney Mardi Gras are so popular amongst heteronormative tourists. Tourists are physically separated from the gay bodies on parade. When spatial segregation is maintained, there can be no confusion between heterosexual and homosexual bodies. The threat of sexualised transgression is, at one level, controlled. The dominant group (heterosexual tourists) can keep their distance from the Other.

At the parades that I attended road barriers were erected along the sides of the streets. The barriers created a wide space in the middle of the streets which tourists could not access. At the 1996 Sydney Mardi Gras the barriers were extensive and formidable (see Figure 4.1). Metal frame crowd control barriers, which were approximately one metre high, stretched the entire length of the parade route. The barriers were supplied by the Sydney City Council and were fixed into place several hours before the parade started. Streets were closed to traffic from approximately 3.00 p.m., five hours before the start of the parade.

Safety first

The *Sydney Star Observer*'s (1996) guide to the 1996 Mardi Gras festival, parade and party, reported that:

> The Parade begins at around 8 pm at the corner of Liverpool and Elizabeth Street, moves up to Oxford Street, turns right into Flinders Street

Street scenes 59

Figure 4.1 Roadside barriers at the Sydney Gay and Lesbian Mardi Gras Parade (source: Author's own collection).

and then left into Moore Park Road. As vantage points in Oxford Street are usually the first taken, try Flinders Street or Moore Park Road. Remember that after 7 pm it gets very difficult to get a good possie anywhere.

The attention paid to places from which to view the parades, or to getting a 'good possie', is imbued with discourses of safety. The numbers attending the Sydney Mardi Gras have increased dramatically since the parade's beginnings. The *Sydney Star Observer* (1996) guide reports on some of the historical changes:

> But how things have changed! From 1981 ... Sydney summer and Mardi Gras energy have seen crowd numbers and float numbers grow. Some of us remember when there were no barricades between the crowd and the Parade and you could jump in and out of the marching throng at will. Don't try that now. It's – after all – safety before spontaneity these days. Crowds in 1994 topped 600,000 ... The float count should be well over 100, there'll be several thousand participants, and more than 800 officials of various kinds.

The '800 officials of various kinds' are usually volunteers who wear uniforms that distinguish them from spectators and paraders. The parade officials carry radio-telephones, hand-held megaphones and whistles. In 1996 I waited for three hours behind the barriers at Oxford Street before the parade began, and during this time, I noted several things about safety procedures. If tourists wished to cross the street they had to ask the officials to let them through the barricades and over the street. During the parade there were no opportunities to step out from behind the barricades and cross the street. If spectators attempted this, they were stopped and encouraged to remain behind the barricades.

Similarly, at HERO parades the crowd was encouraged to stay behind the erected street barriers. Auckland crowds were disciplined before and during the parade. One of my duties, when I worked at the HERO Parade workshop in January and February 1996, was to find 30 volunteers who could be placed on streets that intersected with Ponsonby Road. Each volunteer had a 'marshal' T-shirt, reflector vest and distress flares. In addition to these marshals, each parade entry had to provide two of their own marshals to walk beside their float and to maintain an 'appropriate' distance between paraders and tourists where there were gaps in the road barriers. Marshals were briefed at a pre-parade safety meeting with the organisers and with police. The use of road barriers at the HERO Parade has increased each year with the size of the attending crowd.

The 1997 and 1998 HERO Parade organisers made use of the Auckland City Council's road barriers. These are large plastic 'container type' barricades which, once in place, were filled with water. These barricades were wide and approximately waist height. Marshals and police kept the crowd behind the barricades. Tourists' bodies, for the most part, become disciplined, controlled and carefully separated from the homosexual bodies on parade. By extension, paraders' bodies were also disciplined, controlled and contained within the barricades.

There are several implications of this attention to 'crowd' safety. In making the parade site 'safe' and controlled, a distinction is created between paraders as spectacle and heterosexual tourists as watchers. When safety is well publicised and barriers are erected and maintained by barricades, marshals, and police, more people are attracted to the parade site. Of course one could argue that the barriers *really* are there for safety, to stop the crowd being crushed by a large truck, or spectators run over by motorcycles. My experience, however, from attending parades is that a barrier tended to *increase* risk of physical injury. The barriers concentrate large numbers of people in small spaces and because people cannot move away from the roadside barriers they, ironically, become dangerous obstacles.

At the 1996 HERO Parade, I was elated to see the final parade 'product', which I had helped to create. I was also video recording the parade and hence did not form part of the main crowd at the side of the road. I walked with the parade at times. Carrying a video recording camera marked me as different from the other spectators. At 1997 HERO Parade, however, when I observed the parade from behind the barricades as a lesbian tourist, surrounded by mostly heterosexual tourists, I felt uneasy. I was physically 'trapped' and I became caught up in the feelings of those around me. In field notes made after the parade, I reflected on my feelings of being a tourist at the 1997 HERO Parade and I documented some of the comments that were made while I watched:

> One woman said to the man who was watching the parade with her: 'There are some *normal* people in the parade, there are some *normal* people in the parade, you know, *straight* people. It's not all gay.'
> 'Oh – *what?* – that's sexually dysfunctional' (said a man as he watched a float containing a lesbian sadomasochistic performance).
> (Participant observation notes, February 1997)

These comments can be interpreted as heteronormative tourists' attempts to reconstruct gay bodies as deviant and 'Other', and establish their own 'normality'. In addition to the physical barriers that maintained some distance between parading bodies and tourists, some people also employed other measures to preserve the border between self and Other. For example I observed 'coupling' of heterosexuals at all parades: men draped their arms around 'their' female partners; women and men held hands; and, even more provocatively, some engaged in kissing and other sexual behaviour in the street. Stereotypical jokes were made by heterosexual men about 'keeping their backs to the wall'. I asked a Pākehā heterosexual man: 'Why did you come to the parade tonight?' He responded: 'Because cute women hang out with gay boys.' Such a response serves to maintain distance between the self and the Other, the tourists and the hosts.

Queer enclaves

The majority of tourists at HERO and at the Sydney Mardi Gras are, as I have already noted, heterosexual. There are, however, also gay, lesbian, bisexual and transgendered tourists behind the barriers. At the 1996 HERO Parade I was aware that the spaces occupied by queer tourists were quite different from the spaces occupied by heterosexual tourists. I began distributing questionnaires several hours before the parade was due to start (that is, around seven o'clock that evening). Doing this on Ponsonby Road, which has many gay-owned and operated restaurants and bars, I became aware that many of the people eating at the tables lining the street were gay. They were very willing to fill in my questionnaire, talk to me about my research, and openly identify as gay (the majority were gay men). They had positioned themselves in the restaurant seats by the road, as well as upstairs in restaurant and bar windows, and had come to watch the parade and engage in associated activities ('find a man', 'check out talent', 'watch the "girls"'). In areas like this, gay tourists maintained spatial enclaves away from the large number of heterosexual tourists. Gay guides to viewing the

Sydney Mardi Gras Parade suggested that people could hire rooms in hotels, guest houses and restaurants from which to view the parade.

The pleasure obtained by these queer tourists provokes a mobilisation of the definition 'tourist'. On the one hand, queer tourists can be positioned as hosts and part of the gay pride parade spectacle for non-queer tourists. Gays and lesbians watching the parade, eating at restaurants, drinking at gay bars, and 'picking up' partners on Ponsonby Road are in their 'authentic' or home location (Podmore 2001). On the other hand, some queer tourists position themselves as 'normal' or 'ordinary and everyday' (self), and position the gay paraders as exotic and extraordinary (Other). These binaries of self/Other, straight/gay and tourist/host are subject to contestation and mobilization; they are never stable or static.

A fenced-off area containing tiered seating created another bordered area at the 1998 HERO Parade. This was a fundraising initiative by HIV/AIDS organisations with pre-sold tickets at $45, which is expensive given that many spectators are attracted to the parade because it is free. The area included a bar where drinks could be bought. The majority of people in the tiered seat section were queer. Couples could visibly 'be' together, hold hands and so forth. I could not hear any comments that may have been made by heterosexual tourists which Othered and/or degraded bodies on parade.

While I had thought this space would be a 'queer spectator enclave', it also contained a complex mix of 'VIPs' such as the then Prime Minister of Aotearoa/New Zealand, Jenny Shipley (accompanied by her husband), other Opposition politicians and media representatives. The VIPs were in an area demarcated with a white picket fence, while queer tourists sat on the tiered seats beyond. The white picket fence seemed to act like a sanitised border which kept the 'extraordinary' people and the 'ordinary' queers separated. I sat at a spare table in the VIP area. In this area 'we' – queers, straight and gay politicians, queer entertainers and television celebrities – became another part of the tourist entertainment. Television cameras focused on the Prime Minister's reaction to each float. People from outside of the area watched us. I found *myself* watching for the Prime Minister's reactions as floats passed by her. My roles as researcher and queer activist were overshadowed by the allure of celebrity watching! Behind 'us', in the tiered seating, gay men made sexual comments about the Prime Minister's husband. Thus, discursive and material borders between straight/gay, self/Other, tourist/host shifted constantly.

The physical, or material, borders that operate at HERO and the Sydney Mardi Gras are predicated on discursive borders between paraders and tourists. In the next section I consider psychoanalytic theory on abjection (see Kristeva 1982; Douglas 1980; Grosz 1989, 1994; I. M. Young 1990) as a way of understanding further some of the reasons why HERO and Mardi Gras have become such popular tourist events.

Bordering on abjection

I. M. Young (1990) uses the concept of cultural imperialism to examine general forms of group oppression and violence. Cultural imperialism works to keep a group invisible at the same time as it is marked out and stereotyped. The most visible Others, for example, women, Blacks, disabled people, can be clearly marked out as different from the dominant white, male subject. A border anxiety is present, however, when the Other is least visible. I. M. Young (1990: 146) argues that 'homophobia is one of the deepest fears of difference precisely because the border between gay and straight is constructed as the most permeable; anyone at all can become gay'. The border is most threatening when the gay body cannot be distinguished from the straight body. Only when gay bodies are clearly marked as different, as in gay pride parades, does this border become visible and therefore less threatening to the dominant culture.

This corporeal border was illustrated in many questionnaire responses. At the roadside (behind the barrier) one respondent (male, white and heterosexual) had come to the HERO parade to 'have a look', and defined the HERO parade as a tourist event because 'It's strange, a freak show and a laugh if you're straight'. This response can be read not only as an attempt to mark the parade participants as different from the dominant white, heterosexual subject, but also to mark the parade bodies as 'freaks'. Several responses from heterosexual tourists (including men and women of differing ages and backgrounds) illustrated a desire to maintain a border between straight and gay:

> 'Alternos deserve to have their lifestyles exposed a bit – makes us more comfortable'
> 'To watch the strange people'
> 'To perve'
> 'Have a look, entertainment'
> 'Out of curiosity'
> 'Have a look, to see it first hand'

There were many floats that directly emphasised the threat that queerness poses to social order. For example, the 'Demon Float' and 'Salon Kitty' (which is a bondage and sadomasochistic float described as 'Rubber/Latex fetish: Dressing for pleasure'), both challenged the heteronormative notions of acceptable sexual desire and pleasure through their displays of sadomasochism; the 'New Zealand Prostitutes Collective' brought the bodies of illicit sex workers into public view; the TransPride float challenged the authenticity of rhetoric about two genders/sexes; 'Miss Kitty and Friends', which consisted of a six foot six tall drag queen with two men on dog leads dressed as poodles, provided an animated debate about sexuality, bestiality and gender/sex roles; the 'Safe Sex: No Ifs, No Butts' float consisting of a large revolving polystyrene penis and eight dancing men and two women whipping each other exposed the penis/phallus in a public place; and finally, the 'Body Positive' (an HIV/AIDS organisation) and the Herne Bay House floats, brought defiled, diseased bodies into view. Such bodies do not constitute proper social bodies (Grosz 1994). They threaten to disrupt order and purity, but, and at the same time, they reinforce societal order by remaining in the parade and not spilling into the watching tourists. Thus, these parade entries, which tended to be perceived as some of the most risqué, also tend to reinforce a dichotomy between heterosexual and homosexual.

Julia Kristeva's (1982) notion of abjection is very useful in understanding the ways tourists at gay pride parades combine fascination in, with revulsion against, queer bodies. Abjection 'is a desire for separation, for becoming autonomous and also the feeling of impossibility of doing so' (Kristeva 1982: 135), and Kristeva argues that the abject provokes fear and loathing because it exposes the borders between self and Other as fragile, and threatens to dissolve the subject by dissolving the border.

Drawing on Kristeva's work, Grosz (1994: 193) differentiates between types of abjections: 'abjection towards food and thus toward bodily incorporation; abjection toward bodily waste, which reaches its extreme in the horror of the corpse; and abjection towards signs of sexual difference'. Different floats in the HERO Parade can be viewed as illustrative of each of these categories. The first category, abjection towards food and bodily incorporation, can be linked to paraders who reject the notions that:

> A slim, fit body is for some a source of pride to be paraded in public places, spelling discipline, success and conformity, whereas fat is seen as a sign of moral and physical decay. Fat people are stereotyped as undisciplined, self-indulgent, unhealthy, lazy, untrustworthy, unwilling and non-conforming ... Unlike the disciplined slim body the fat body is not welcome in everyday places.
>
> (Bell and Valentine 1997: 35–36)

Few lesbian bodies on parade represent existing norms of 'feeding regimes'. Many of the lesbian bodies on display are large. Paraders, such as Dykes on Bikes, Marching Girls, and the lesbian float called Lassoo, tend to ignore disciplinary regimes that aim at making a slim body. Instead there is often pride associated with being large. Tourists' abject reactions to large lesbians are evident in some of their comments. For example, two men exchanged the following comment about a large Dyke on a Bike: 'Oh, God, did you see her? She's *huge* [pause] nice bike though.'

The second category of abjection, bodily waste and horror of the corpse, is integral to many floats at the HERO Parade and Mardi Gras. Both parades are organised to raise funds for HIV/AIDS organisations, and in resting on gay male bodies, they also rest on the notion of abjection. 'Sexuality has become reinvested with notions of contagion and death, of danger and purity, as a consequence of the AIDS crisis' (Grosz 1994: 193). Gay male sex is understood (by 'straight' society) to be predicated on oral penetration and faecal contamination and this carries an unspeakable connection to excremental pollution. Kristeva (1982: 71) argues: 'Excrement and its equivalents (decay, infection, disease, corpse, etc.) stand for the danger to identity that comes from without: the ego threatened by the non-ego, society threatened by its outside, life by death.' Several floats in the HERO Parade can be connected to this understanding of abjection. Specifically, floats which represent HIV/AIDS organisations, such as the Safe Sex float, Herne Bay House float (residential care for people living with HIV/AIDs), Body Positive float, and the Remembrance float, intensify the abject thoughts, that disease is picked up off rectal walls and that death follows from this disease.

Gay pride parades are also significant sites for analyses of sexual difference, the third category of abjection. There are many HERO floats that fit this category. In particular, the following floats and paraders most obviously focused on sexual difference: Drag Queens, Te Waka Awhina (a waka – canoe – with gay and transgender Māori), Mika (singer/entertainer in a queer Māori temple), Sisters of Perpetual Indulgence (gay men as nuns), TransPride (various people from the transgender community), Lost Grannies (men dressed as 'grannies') and Surrender Dorothy (a large shoe and yellow brick road with drag queens). These paraders disrupted and subverted conventional and hegemonic notions of sexual difference. I conducted a focus group with five people who constructed the Transpride float for the 1996 HERO Parade, the theme of which was 'Heavenly', their objective being to upset the construction of transgenders as, what I. M. Young (1990: 123) terms 'ugly bodies'. Aroha began with a description of their float:

Aroha: All the, the costumes on the float, the majority on the float are pastel colours and gold.
Lynda: Great.
Aroha: We wanted to create a heavenly (//)
Chris: (//) yeah yeah.
Aroha: approach.
Janet: Cos we can be pure just like anybody else ... why are we, why are we suddenly, why are we suddenly, considered dirty? (Yeah) because we want to cross a borderline sexually. You know and um, why should we look like sluts when we don't feel like sluts?
Chris: Yeah, get rid of the typecast.
Lynda: Yeah, take away the stereotype.
Janet: Most people think transgender is a mockery.
Aroha: They do.
Janet: You know sort of taking off something.
Aroha: But then when you look at it, a lot of it, public attention is focused on girls on the street (Mmm). They're the ones they see, but it's not always the case.

Lynda: No.
Janet: And a lot of the girls on the street are just making a living, there's nothing else they can do.
Aroha: And there is an awful lot of talent within the transgendered community.

(Transpride Focus Group interview, 15 February 1996)

Aroha, Janet and Chris wanted to challenge the dominant discourses that degrade and debase transgenders as 'dirty', and as 'sluts'. They wanted to do this by invoking the opposite of being defined as ugly, hence their float had a very 'feminine' and 'pure' theme. It was constructed around traditional markers of femininity in terms of colours (pastels and gold), costumes, and other props, in an attempt to offer a transgendered subjectivity other than that of 'working girls on the street'. It could be argued that transgenders represent a type of intolerable sexual ambiguity (Kristeva 1982). The heterosexual (cultural imperialist) tourist constructs a conceptual limit of human subjectivity. Categorisations of subjects as 'freaks', 'ugly', and 'dirty sluts' are attempts to reposition the border between the self and the Other.

The discursive context of pride parades is that they are frequently and predominantly understood as protests for equal rights, or equality. David Smith (1994: 49) argues: 'inequality can be thought of as a particular type of difference between people, about which moral questions arise. Social justice is concerned with this sort of difference.' Smith (1994) refers to a specific, socially constructed conception of normal and acceptable behaviour. I. M. Young (1990: 134) adds: 'when public morality is committed to principles of equal treatment and equal worth for all persons, public morality requires that judgements about the superiority or inferiority of persons be made on an individual basis according to individual competence.'

In many societies there are broad commitments to equal rights and equal treatment for all persons, whatever their group identification. I. M. Young (1990) identifies this as a discursive commitment to equality. She states that: 'racism, sexism, homophobia, ageism, and ableism ... have not disappeared with that commitment, but have gone underground, dwelling in everyday habits and cultural meanings of which people are for the most part unaware' (I. M. Young 1990: 124).

Overt group oppression has, in many western societies that are committed to equal rights, resurfaced as liberal humanism. Liberal humanism treats each person as an individual, ostensibly ignoring differences of race, ethnicity, sex, religion, and sexuality. Structural patterns of group oppression remain and are often unidentified in the rhetoric of equality that liberalism sustains (I. M. Young 1990). The construction of the dominant culture as the norm remains unchanged.

Liberal humanism, or commitments to equal rights for individuals, were evident in the questionnaire responses. Many people at HERO and Sydney

Mardi Gras stated that they thought gay pride parades made Auckland and Sydney 'liberal, contemporary and tolerant' cities. First, I situate these comments within I. M. Young's (1990) theory that dominant discourses of equality and liberty create blindness to difference. Second, I argue that within the parade context, heterosexual tourists are positioned as the dominant cultural imperialist group, that is, as unified, unmarked and neutral.

'Tolerant', 'liberal' and 'open-minded'

At the HERO Parade I asked parade watchers: 'What do you think the parade does for Auckland's image?' At the Sydney Mardi Gras I asked parade watchers the same question in relation to Sydney. Tourists, who identified themselves as heterosexual, responded with words and phrases such as 'tolerant', 'liberal' and 'open-minded'. These responses can be read as underpinned by a neo-liberal humanist notion that society is composed of individuals who have commitments to their own autonomy, the general idea of liberty, and the notion that this liberty constitutes the primary social good. Group difference and the lack of liberty for certain group memberships, nevertheless continue to exist. Certain dominant groups are privileged and other groups have their liberty consequently compromised (I. M. Young 1990). Insisting that all individuals are equal entails ignoring difference, which has oppressive consequences.

Responses from Auckland and Sydney include:

'General acceptance/tolerance'
'Increases tolerance and awareness'
'Open-minded city'
'Cosmopolitan, open-minded, liberal'
'Gives a party/tolerant image'
'Shows broadness of mind'
'It shows that we are more liberal than other cities, progressive and modern'
'It shows that we are open-minded and tolerant city'
'Makes us more open to everyone'
'Makes us more enlightened, open-minded'
'It tells other cities how liberal we are'
'Makes us open to all walks of life'
'Great – shows how open-minded people in Auckland are'
'Shocks people, but encourages open-mindedness'
'Positive – makes people open their minds'
'Shows that Auckland has a liberal population'
'Positive, lively, diverse, tolerant, fun'
'It makes it funnier, it seems more open-minded than other cities'
'The people of Auckland are hopefully more tolerant of others'

'Shows a tolerant attitude towards gays and sexuality'
'It definitely portrays it as a liberal (in terms of American liberals) city'
'Promotes Sydney as a friendly city and show that gays and lesbians can definitely fit into Hetro [*sic*] society'

Blindness to difference disadvantages groups whose experience, culture and socialised capacities differ from those privileged groups ... The strategy of assimilation aims to bring formerly excluded groups into the mainstream (I. M. Young 1990: 164). The ideal of a universal humanity without group differences allows privileged groups to ignore their own group specificity. The 'heterosexual' responses of tolerance, liberalness, and open-mindedness, however, mark the bodies on parade as different and as Other from the dominant and normalised social group, heterosexual tourists. This group seem to have to constantly remind themselves to be 'tolerant' and 'open-minded' in relation to the Other.

The *Compact Oxford English Dictionary* (1991: 2075) defines tolerance as:

> The action of allowing; licence permission granted by an authority ... The action or practice of tolerating; toleration; the disposition to be patient with or indulgent to the opinion of others; freedom from bigotry or undue severity in judging the conduct of others; forbearance; catholicity of spirit.

Being tolerant, therefore, relies on there being an Other, without any examination of the self as the dominant group. The dominant group remains the norm.

The responses also show that there is an unstated 'us' at work here. One of the comments is: 'gays and lesbians can definitely fit into Hetro [*sic*] society.' This comment attests to I. M. Young's (1990) notion of assimilation. The 'strategy of assimilation always implies coming to the game after it has already begun, after the rules and standards have already been set, and having to prove oneself according to which all will be measured' (I. M. Young 1990: 164). At the HERO Parade and the Sydney Mardi Gras, heterosexual tourists can be identified as the 'game starters'. As cultural imperialists they have the rules and standards by which gays and lesbians are judged. Such judgement rests on the idea that queers have to 'fit' into heterosexual society.

The (un)marked tourist

I have been arguing that the bodies on parade become the marked Other and are deviant from the dominant imperialist group, the tourists. There were several indicators on questionnaires that highlighted heterosexuals' privileged position as 'natural' and 'normal'. One indicator may be found in the propensity of many respondents to misspell 'heterosexual'. The questionnaire had several categories of sexuality to choose from. Of those respondents claiming heterosexual status, the majority circled the heterosexual category. Of the 45 people who wrote the word, 19 (42 per cent) wrote either 'Hetro', or 'Hetrosexual'. One explanation for the misspelling of 'heterosexual' could be found in an examination of Aotearoa/New Zealand and Australian accents. Another explanation, however, is to assume that heterosexuals are not often asked to think about their sexuality, state their sexuality, or spell their category of sexuality. It could be argued that the status of heterosexuality is a taken for granted norm. There is another important methodological point to make here. This type of data and subsequent analysis would have been absent (not textually represented) if I – when administering the questionnaires – had verbally asked each question and written the responses myself. My initial reasoning for asking respondents to fill in their own questionnaire was based on an expectation that I would be able to capture more respondents in the limited time available. By asking respondents to fill in the questionnaire themselves, I was able to get groups of people in one approach. I also wanted to make the data collecting process as 'fuss free' as possible, as I had six other people working on my behalf.

The unmarked tourist appeared in Sydney also. The confusion was not so much over the spelling of heterosexual (although this happened also), but centred on the categorisation of ethnicity. In my questionnaire for the HERO Parade, I relied on categorisation for ethnic groups used in the Aotearoa/New Zealand census. In Australia, the census does not require its citizens to distinguish their ethnicity, but citizens are asked in which country they were born, and if they are Aboriginal. I left the ethnicity question open (please state). Many 'white' Australians did not understand the question. Two people, whilst reading their questionnaires asked each other: 'What's ethnicity'? The other one looked puzzled and then replied: 'Oh, she means "authenticity".' People who wrote 'Chinese', 'Asian (Indonesian)', 'Italian', or 'Indian', did not seem to be troubled by this question. 'White' people, however, wrote 'Australian', 'Aussi' or even 'New South Wales.'

Disrupting borders

Responses that exemplified support and pride, especially from those tourists who identified as gay, disrupted the notion of the unmarked tourist. Gay tourists can be conceived of as both self (being subsumed as part of the largely heterosexual audience) and Other (willing to identify as not heterosexual). They may also be conceived as both tourist and host. These respondents tended to embrace and celebrate the parade as part of their identity, rather than attempt to create a conceptual barrier between the paraders and the tourists. Furthermore, there did not seem to be any misunderstandings over questions, gay respondents did not make fun of the questionnaire, and no one misspelt their sexual category (for example lesbian, gay, queer, transgendered, dyke and so on). Some examples of responses to the parades are:

'To support it and have a look and suss out talent.'
'Promotes Auckland as a gay oriented city.'
'Improves it. Makes Auckland cosmopolitan. Shows that there are many different groups in New Zealand and they should all be represented. Many people are not Christian.'
'To be part of some thing I am and that I don't have to hide behind.'
'Because I am gay and very supportive of the community I live and work in.'
'Living and working in this gayest suburb is very important to me.'
'Fun/diversity of inhabitants.'
'Shows that we are a multi-sexual culture! It helps to show those against us that we are "normal", from all walks of life and not necessarily stereotypically "queer".'
'It shows that all groups make up the city.'
'To participate in the occasion, do my part to draw attention to our community, and of course, as a social event!'
'Well, well, well, it's a happening fantastic place to be out and proud.'
'It is an event where you can be yourself, express yourself – FREEDOM.'
'Because Mardi Gras is for queers what Xmas is for heterosexuals.'
'To celebrate what I am and to find a man. To have fun and be happy.'
'To grasp the wonderful atmosphere – to be a proud Australian – to be part of a great moment.'

In the remainder of this chapter I discuss three particular floats of the 1996 HERO Parade and examine the ways in which each one discursively 'disrupts borders'.

Rainbow Youth organisation

I conducted a focus group with Rainbow Youth at the HERO Parade workshop and began by asking about the Rainbow Youth organisation.

Rachel: Um, we are an organisation ... um that provides services for, um, gay and lesbian and bisexual people.
Lynda: Yes.
Rachel: And we do things like school workshops, um, run a social group, a women's one called 'Dykelets and Bykelets'
Rachel: And we went on a confidence course out at [name of high school]
Nick: And we provide a regular newsletter.
(Rainbow Youth Focus Group interview, 14 February 1996)

I was interested in their work in high schools, so I questioned them further. I began with an assertion, that:

Lynda:	Nobody wants to know about queer people in their classrooms. Have you had any, like, backlash or ?(//)
Rachel:	(//) Not in a big way. That [name of School Principal] guy said there were no queer people in his school. And Simon is from his school [turn and look at Simon] [All laughing]
Lynda:	And, and, who is this guy?
Nick:	The Principal and an Auckland City Councillor.
Lynda:	Oh no. He said there was no (//)
Rachel/Nick/Sam:	Yes.
Nick:	I might add that actually at that time, Sam was actually in one of our groups.

(Rainbow Youth Focus Group interview,
14 February 1996, emphasis in original)

I have included this exchange as it illustrates the way that some high school authorities (and Auckland City Councillors) try to deny the existence of young, gay, lesbian, and bisexual students. The only way the school Principal can defend his heterosexual identity is to turn away with irrational disgust. The Rainbow Youth parade entry built on this denial of their existence in a way that undermines and jeopardises abject reactions.

The Rainbow float contained young people dressed in various Auckland high school uniforms. Their banner read: 'We're Here, We're Queer, We're in Your Classroom'. Their school uniforms marked their bodies as self, not as Other, as mind, not as body. Tourists at the HERO Parade, intent on consuming the Other, have this desire disrupted. Rainbow Youth discussed their parade objectives in a focus group interview.

Rachel:	I think that's why, it's basically why I think it's a good idea. We're getting out there and part of our, our float is that we're all dressed in school uniforms and our banners like 'We're Here, We're Queer, We're in Your Classroom', kind of thing and it . . . and it lets people know that, that we are in the classrooms and also for people who might be watching who are queer, it's quite good to know that there are other people out there going through the same thing.
Lynda:	Yeah. A lot of the people going in the parade dressed up to be really outrageous. So it's quite ironic that you're putting on, um, you know uniforms to be like regular life out there. It's a really nice kind of twist to the whole parade.
Nick:	The other thing is that the very fact that we are wearing uniforms is in its own way outrageous . . . In some ways, um, the point, the point of our float is to shock people a little bit . . . It's to wake them up and say 'we're here' and in that respect it

is going to be quite outrageous ... Um, it'll be interesting to *see people's reactions when they see their school uniform.*
(Rainbow Youth Focus Group interview, 14 February 1996, emphasis added)

Nick's remark highlights the possible border anxiety of the self/Other relationship between parade participants of the Rainbow Youth float and the roadside tourists. A float representing the self, such as the Rainbow Youth float, upsets and disrupts this dichotomy. Tourists watching may recognise their – or one of their children's – school uniforms. As a consequence, the border between straight/gay, mind/body and self/Other becomes blurred. Heterosexual tourists are faced with a disruption of their 'normalised', unmarked identity.

There were other groups involved in the HERO Parade, directly and indirectly, that also destabilised the mind/body and self/Other dualism. One such group was Gaily Normal.

Gaily Normal

The name Gaily Normal establishes the possibility for border anxiety. 'Normality' is usually equated with heterosexuality (this became evident in the questionnaire responses). Gaily Normal are a political gay group that attempt to construct themselves as self, rather than as deviant Other. They object to all homosexuals being stereotyped as, for example, drag queens, marching boys, or as involved in s/m sexual practices.

The *New Zealand Listener* (Gearing 1994: 26–27) published an article, 'Hero V Homebody: a gay debate', in which Gaily Normal's objections to the HERO Parade were articulated.

> Gaily Normal ... announce the sound of a new voice in the gay community, after a call by the large number of people who do not feel heard or represented under current circumstances. This has become particularly evident with the recent controversial issue of simulated sexual and violent explicitness in parts of the Hero parade. Many gay people are not willing any longer to be 'lumped in' under the always most visible umbrella of drag, S & M, and other generally explicit content. A large proportion of the gay community live lives not dissimilar from most 'straight' people.

Gaily Normal also added 'It is time for gays to blend in with heteros' (quoted by Gearing 1994: 26). In this instance, Gaily Normal attempt to align themselves with the dominant discourse of heterosexuality, or the heterocentric mainstream. This caused a backlash from other members of Auckland's gay community. A HERO Parade organiser (quoted by Gearing 1994: 27) reacted to Gaily Normal in the following way:

I am not prepared to have our community present a sterile version of itself to gain acceptance from the alter. If that was the thinking then we wouldn't be this far down the track. There are Doctors and Lawyers on the floats, anyway. Do you want to see a float of two gay men reading the paper and drawing up a Foodtown [supermarket] list? Normal, but hardly exciting. Thank God that we have the edge that excites.

The HERO Director (quoted by Gearing 1994: 27) claimed that Gaily Normal were a symbol of self-oppression by stating: 'It is a trend in all minorities struggling for identity in a suffocating world ... Others in this minority, desperate to be part of the dominant culture, will do anything to maintain acceptance or normality.' This reaction to Gaily Normal, by the HERO project, further reinforces a straight/gay binary. Gaily Normal is constructed by the HERO project as politically ineffectual and as passively cohabiting with a hegemonic, heterosexual culture, which is oppressive to sexual dissidents.

Gaily Normal can also be read through I. M. Young's (1990) argument about cultural imperialism. I have already stated that when the homosexual body cannot be told apart from the heterosexual body, then the border anxiety is greatest. Gaily Normal can be understood as part of 'normal' society and hence, more threatening to cultural imperialists.

Robert discussed with me the conflicts that emerged between Gaily Normal and his parade entry, the 'Demon Float'.

Robert: Gaily normal were saying 'we don't want these freaks showing themselves in public, there are gay people who are normal'. We kind of got really pissed off with that because, I mean, for me there are, for me there is lot of difference within the gay community ... And why shouldn't we represent the whole side of it ... The whole spectrum?

Lynda: Do you think that it stereotypes, that the HERO Parade stereotypes, some of the gay community?

Robert: Oh hideously, yeah. But then the gay community is very conservative in New Zealand anyway. It's quite um, frightened of difference. It is actually quite conservative, that, that once you start to look different or stand out, you aren't accepted as readily and you actually get a bit of a hard time.

Lynda: So there seems to be like a code of behaviour?

Robert: Yeah.

(In-depth interview, 7 February 1996)

Robert's analysis is indicative of several conflicting assumptions surrounding the HERO Parade. On one hand, the HERO Parade produces a proliferation of difference (there is a lot of difference in the gay community). On the other hand, the HERO Parade stereotypes bodies of the parade as the unified

Other of the (mostly) heterosexual audience. In relation to I. M. Young's (1990) argument, Gaily Normal can be read as very transgressive because they threaten to spill into the ordered society, almost unnoticed, and therefore, undetectable.

Gaily Normal provided an alternative representation of the 'gay community' and hence a disruption to the dichotomies of self and Other, straight and gay. The Gay Auckland Business Association (GABA) also disrupted these binaries.

Gay Auckland Business Association

The Gay Auckland Business Association (GABA) entered a float in the 1996 HERO Parade. Their float was an 'office scene' with people dressed in business suits, working at computers and answering phones. It was representative of what one participant described as a 'professional' image.

Justin: GABA is more a business organisation that, um, it's there for networking purposes ... It has quite a low profile and that's the thing that we are trying to change with GABA because not very many people know of it, certainly a lot of gay people don't. So, as far as getting involved in gay marches or events or whatever, GABA has actually been very low key.

Lynda: Right.

Justin: Um, one, because a lot of gay professional people are quite reserved about being gay. They are not so out during the day or visibly [out]. And secondly we haven't had a lot of um, what's the words, support, flags, bunting whatever to say 'hey, here we are'. So the first one, really, that we were visible in, was the recent Coming Out Day Parade where we did have a float. That's the first one for quite a while.

Lynda: And you were giving out fliers as well?

Justin: Yes. So we had a float made up. It wasn't spectacular by any means but it costs money and that's basically where it all comes back to. The cost of putting a float in. And we simply ran that down Ponsonby Road and we had three or four of us handing out leaflets on GABA, letting people know because, again, to let people know is difficult.

(Individual interview, 24 January 1996)

Justin claims that many gay professionals are quite reserved and that they are not 'out' during the day. One of the reasons for this, I believe, is that there are many risks involved in being an 'out professional'. Once 'out(ed)' cultural imperialists may 'turn away in disgust and revulsion' (I. M. Young 1990: 146). Being a professional also means occupying a position which is conceptually aligned with the mind. If the border between mind and body is crossed in a 'mind' job, then the fears of the dominant group are unleashed. Justin also invokes day/night and serious/fun binaries for professionals. A day time job is most likely to be understood as professional. A night time job is more likely to be conceptualised as deviant and aligned to the body.

A well-known and outspoken Aotearoa/New Zealand journalist, Warwick Rogers, who claims he is not homophobic, sums up this dichotomy:

> And I'm in two minds about gays who prance half-naked along Ponsonby Road on Saturday night in February and then expect to be taken seriously as doctors, lawyers, accountants and teachers the following Monday morning. But good luck to them, I say, if prancing makes them happy.
>
> (Rogers 1998: 6)

Doctors, lawyers, accountants and teachers are all 'professionals' and, according to Rogers (1998), should not be both of 'the body' and of 'the mind'. If a gay professional is not 'out' in the work place, then their sexuality is usually assumed to be heterosexual. In the HERO Parade, 'out' professionals upset this connection between heterosexuality and professionalism. Rogers (1998) could be described as a 'professional' journalist. He attempts to distance himself from the 'professional queer bodies' because they threaten to split and disrupt his heteronormative identity.

Conclusion

Although the HERO and Mardi Gras Parades contest the everyday construction of streets as heterosexual, territorial strategies of containment and control of homosexuality are fundamental to the success of the parades. They therefore provide important sites from which to discuss power relations involved in tourism processes. At these parades, borders are maintained between tourists and hosts, which are crucial to their success as spectacular events. The physical or material borders keep the (largely heterosexual) crowd separated from the queer bodies on parade, and, for the watching heterosexual tourists, this physical separation takes some of the 'threat' out of homosexual bodies. Simultaneously, a conceptual border separates those who are perceived to represent the body, from those who are perceived to represent the mind. This mind/body dichotomy becomes aligned with heterosexual tourists/homosexual hosts.

I. M. Young's (1990) theory of cultural imperialism adds weight to my argument that queer bodies on parade become 'othered' by the heterosexual tourists. Straight tourists, or 'cultural imperialists', culturally inscribe homosexual bodies as deviant, freaks and as ugly, by which they are both fascinated and repelled in ways consistent with Kristeva's notion of the abject.

My discussion has involved theoretical and political tensions surrounding notions of liberalism, group identity and difference, and therefore inserts 'political struggle' into tourism studies. Gay pride parades provide an opportunity to deconstruct acts of tourism, pleasure and politics as these are lived through the bodies involved. This discussion of the HERO Parade and the Sydney Mardi Gras, and relationships between heterosexual tourists and gay hosts, shows that such tourist events are both complicit with dualistic mechanisms of western thought, and, at the same time, they contest hierarchical dualisms through a disruption of the cultural imperialist position as 'normal' and neutral. Boundaries are central to western conceptual frameworks of space and bodies. However, as my account has shown, it is possible to mobilise dualisms and to produce contradictory readings and experiences of gay pride parades.

5 Sex in the suburbs or the CBD?

Gay pride parades may queer streets, cities and nations, but not without a considerable amount of controversy. Part of the controversy is that 'public' opposition (and support) to this kind of activist/tourist event rests on an assumption that queer bodies should do 'queer things' in private spaces. Pride parades foreground the problematic – and sometimes eroticised – position of 'private' bodies in public places (see Johnston 1997). In this chapter I discuss the contested terrains of parades and argue that these embodied sites are contingent upon, and mediated through performative western hierarchical dualisms, such as mind/body, straight/queer, public/private. These binaries are central to western thought and have received much attention in geography (Berg 1994; Johnston 2005; Longhurst 1995, 2001a; McDowell 1991; Rose 1993b; Sayer 1989; Vaiou 1992).

Most tourist sites rely on visually spectacular distinctions which demarcate them as extraordinary places and the embodied presence of pride parades enables a particular construction of tourist sites. The site of pride parades do not just use cities as their spatial back drop, rather, cities derive their tourist meaning from the queer bodies that parade in them. It has been argued that spectacular tourist sites are those which 'interfere with ordinary collective life-routines by focusing consciousness on a documented external event' (Rojek 1997: 64). Notions of the 'ordinary' and the 'everyday' are commonly and unproblematically used in literature on tourist events. This taken for granted notion of the 'everyday' often obscures hegemonic investments in 'everyday' meanings. In my research on gay pride parades, I identify that the 'everyday' tends to be socially constructed heterosexuality. Gay pride parades attempt to construct, spatially and temporally, an attractive and unique performance in opposition to the 'everyday', heteronormative spaces of cities. Chris Brickell builds on these notions (2000) in an article entitled 'Heroes and Invaders: gay and lesbian pride parades and public/private distinction in New Zealand media accounts'. He uses media accounts to understand the ways in which discursive inversions occur whereby homosexuals become constructed as powerful and tyrannous, and heterosexuals become understood as being coerced and oppressed. In working through this material I adopt dialogues between gender and sexualised theorisations and performative geographies as these offer opportunities to understand the complexities and mutual reformations of gendered and sexualised bodies and city tourism spaces.

This chapter presents a particular spatial battle that was played out in Auckland city over the site of the parade. I argue that queer bodies on parade are perceived to be inappropriate bodies to inhabit public places such as the CBD of Auckland. Queen Street, the main business street of Auckland, is analogous to the 'mind': the site of rational decision making. The subsequent shift of the parade site to Ponsonby (see Figure 5.1), a 'gay' suburb of Auckland, reinscribes both the site and the bodies as 'private' and queer. Ponsonby Road is analogous to the 'body': the site of consumption, appetite, desire and domestic repose.

The HERO Project employs marketing strategies to create a local gay environment and a local gay identity. The debate over the site of the HERO Parade demonstrates the hegemonic representations of particular streets and particular bodies in Auckland. Hence, ironically, the argument rests on the dichotomous relationships between public/private and heterosexual/homosexual, at the same time that these dichotomies are problematised.

Heteronormative tourists' participation in the parades not only 'successfully' Others the gay pride paraders, but calls the borders of corporeal acceptability into question. This is the subject of the next section. The focus shifts towards the construction of the HERO Parade and the public debate that ensued over the specific place and bodies of the parade in Auckland, Aotearoa/New Zealand.

The inaugural HERO Parade was held on Queen Street, in the 'heart' of Auckland's CBD, on the night of 19 February 1994. The timing was important because it was less than one year after human rights legislation was passed in Aotearoa/New Zealand which made it illegal to discriminate against homosexuals (Gearing 1997). It could be argued that the HERO Parade began as a sort of protest and celebration of the human rights legislation, but the parade has developed over several years in a way that it now has multiple and competing meanings.

Pride parades have the potential to queer cities; they also, however, tend to be caught up in dominant discourses that construct queer bodies as Other, and parade watchers as tourists. The spectacle of the HERO Parade is constituted through binaries of tourist/host, and straight/gay. Queer tourists watching the parade, however, disrupt the binary between straight/gay and tourist/host and I return to this in the next section which discusses the debate over the site of the HERO Parade and suggest this debate is contingent upon private and public notions of bodies and places.

The limits to, and possibilities for, contestatory queer politics become evident in the debate over the site of the HERO Parade. There was both support for, and opposition to, the parade being held in Queen Street, in Ponsonby Road, or at all. This debate can be understood as contingent on the mind/body dualism.

Sex in the suburbs or the CBD? 79

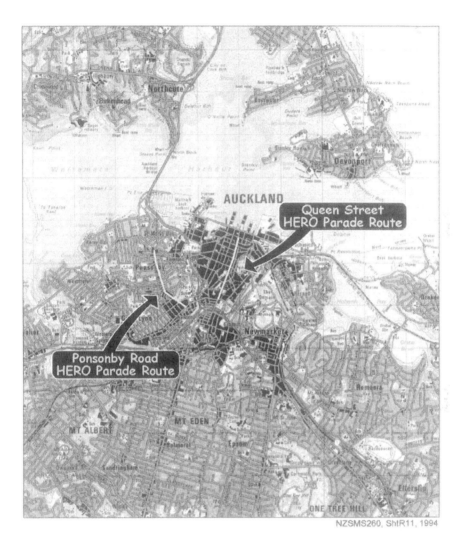

Figure 5.1 Queen Street and Ponsonby Road HERO parade sites (source: Betty-Ann Kamp, Cartographer, Department of Geography, Tourism and Environmental Planning, University of Waikato).

The specific site of the HERO Parade articulates gendered/sexed and sexualised notions of embodiment. The site of the parade is an important focus of this research as it is the contested and constructed terrain which tourists are drawn to. The site of the HERO Parade both confirms and disrupts mind/body, public/private and tourist/hosts binaries. Prior to 1996, the HERO parades had been along Queen Street and through the CBD of Auckland. In 1996 the parade was reassigned to Ponsonby Road, a key commercial street in a suburb, which is often described as Auckland's gay ghetto.

I argue that Queen Street represents the public and straight 'mind' of Auckland, while Ponsonby has become Auckland's private gay 'body'. I draw on newspaper articles, letters to newspaper editors, radio and television transcripts and interviews to support this claim. The public debate over the HERO Parade site served to reinforce binary and hegemonic representations of bodies as either straight or gay.

Queen Street

The first two Queen Street HERO parades of 1994 and 1995 were hotly contested and troubled events. Approximately 400 people participated in the 1994 parade and the crowd watching the parade was estimated at 100,000 (Gearing 1997). There were only seven floats in the first HERO Parade; they included three drag queen acts, an s/m safe sex float and two lesbian floats (Gearing 1997).

Auckland City Councillor and Deputy Mayor, David Hay (quoted in *Sunday Star Times*, 10 April 1994: A5), described some of the 'inappropriate' bodies of the first 1994 Queen Street HERO Parade. 'A whole lot of men that had G-strings on and nothing much else, and bare-topped women, and just a lot of sights that I don't think are suitable for Auckland.'

In a television news programme (*Eyewitness News* 1994) David Hay elaborated on this claim:

> *Presenter:* And so Councillor Hay decided to record the event on his own home video. Councillor, this is a very public place, the middle of Queen Street. Is this what you are objecting to here?
>
> *Hay:* Yes, I am. Queen Street has always been a place for our parades of our heroes – our Olympic team ... the All Blacks have been up Queen Street, the Whitbread parades up Queen Street ... I don't think it's an appropriate place for bare-breasted women and nudity in our Queen Street. I spoke to a number of people that had been offended, had been caught unawares. They weren't aware that it was on. They'd been to visit the Whitbread boats and, in fact, one couple thought that the parade was for the heroes of the Whitbread yachts.

Of significance here is the appropriation of the word 'hero'. Heroes are, according to Deputy Major Hay, New Zealand's international sporting representatives. In particular, New Zealand heroes are from celebrity male sporting events, such as New Zealand's representative rugby team, the All Blacks, and the New Zealand Whitbread yachting team. These New Zealand heroes are the 'right' bodies to parade in Queen Street because they

Sex in the suburbs or the CBD? 81

exist in the dominant national imagination as the 'real' heroes of New Zealand (see for example Bev James and Kay Saville-Smith 1994, and Jock Phillips 1996, for accounts of hegemonic masculinity and nationalism in Aotearoa/New Zealand). In other words, these sporting 'heroes' can be trusted in public places because they do not publicly identify or perform as homosexual, they are not effeminate or feminised in any way, and they usually remain fully clothed. There have been occasions, however, when extremely muscular Aotearoa/New Zealand All Black rugby players have unexpectedly taken their shirts (not pants) off in public to both 'thrill' their spectators and raise money for charities (certainly not HIV/AIDS).

Further complaints against the HERO Parade in Queen Street dominated the letters to editors of Auckland newspapers. One began with 'Horrible Heroes' and stated:

> I PROTEST at the city council funding [*sic*] and allowing the 'Heroes' parade though our city. Admittedly the law allows for no discrimination regarding colour, race, sex, etc., but until homosexual people were allowed 'their right' to 'come out of the cupboard' we never had these kinds of brazen sexual parades. They do nothing to restore decent moral standards at a time of excessive violence and crime and for the council to support this activity is deplorable. We are proud to salute our national and sporting heroes in parades up Queen St but it is a travesty of the word 'hero' to have a parade of this nature. Are these the sort of 'heroes' we want our young New Zealanders to identify with?
> (*Sunday Star Times*, 6 March 1994: 4, capitals in original)

The writer of this letter assumes that the city of Auckland and Queen Street, in particular, belongs, not to homosexuals, but to heterosexuals. He is also offended by the use of the term 'hero'. Heroes are, once again, referred to as 'our' national and sporting heroes. Only these sporting heroes should be allowed to parade in Queen Street, Auckland.

The contradiction, that parades are simultaneously allowed and disavowed, is best understood in terms of public and private. Parades of 'New Zealand's Heroes' are permitted in Auckland's most public and prominent business centre – Queen Street – when they do not transgress hegemonic notions of sexualised embodiment. A Queen Street HERO Parade is challenged by the Mayor, Deputy Mayor and some other Councillors (the 'city fathers') because such a parade endorses the acceptance of 'private' bodies coming out in public places. Public places, it would seem, are for 'proper' heterosexual bodies. In other words, Queen Street can be understood to be Auckland's sanitised 'centrepiece', which promotes a 'clean' metropolitan identity. Queer HERO bodies can be understood as 'dirty', with the potential to mess-up Queen Street.

An Auckland popular magazine *Metro* (Legat 1994) ran a story entitled 'Hay Fever' that neatly summarised the parade site debate. Nicola Legat

(1994: 88) reported: 'When David Hay speaks of moral outrage, he always begins with a little old lady. Little old ladies, the meek, self-effacing grandmothers of the land, are the guardians of middle New Zealand's family values.' Hay's use of the 'little old lady' is designed to focus attention on the appropriate, upright, moral corporeality of aged femininity and propriety.

Hay also ran a campaign to remove the HERO festival magazine from libraries in Auckland, so that the delicate sensibilities of 'little old ladies and their innocent daughters' would be protected (Legat 1994: 88). Legat (1994: 89) sarcastically wrote that Hay:

> was concerned about what was being paid for with the help of his rates. Men displaying their posteriors in the city's main street ... How ghastly it must have been! How disturbing! But he had to go on filming, and he had to give the tape to Channel 2 [Television] for screening on *Eyewitness* the next week so that all New Zealand should know the depths of depravity to which the Hero parade had sunk.

This debate between the Deputy Mayor and HERO raged for several months. Hay reiterated his stance against the HERO Parade in Queen Street during a national Radio New Zealand interview (Hill 1994):

> *Kim Hill:* Let's be specific. What was it you found offensive?
> *David Hay:* I don't think it's appropriate to have women and transvestites with bare breasts in Queen Street. I don't believe it's appropriate to have simulated sex in Queen Street on the back of a float ... I believe it's offensive to have bare buttocks walking up and down Queen Street.

This debate also caught the attention of another prominent radio talkshow host, Leighton Smith of Auckland's 1ZB. Smith (1994) discussed his views – on air – regarding the Queen Street HERO Parade:

> I believe that most people in this country still find the sort of behaviour that gays get up to abhorrent, undesirable and a number of other adjectives I could use. That's the way that most people still think and I believe that it's probably the way most people will always think. However, we have become a tolerant society of any number of attitudes, including this particular one. And the thing about the gay community is this, that having gained a sort of acceptance in society where they can do what they want and they can behave in a manner that they please behind closed doors, they are not satisfied with that. They want to take it further, they want equal rights and in many cases they want extra special rights. Personally I believe David Hay is right. Unfortunately what has happened is this: there is a wedge being driven in society so

that those who are of a tolerant nature are being forced to take a stand on issues that they find unacceptable, but some people want to push. That's the bottom line, unfortunately.

In this lengthy quotation, Smith identifies that gays and lesbians are 'abhorrent and undesirable' and are not the types of bodies to be trusted in public places. This radio host thought gays and lesbians should be behind (private) closed doors (the closet). The universality of Smith's claim rings hollow when put alongside another comment by Deputy Mayor Hay to a different Auckland 1ZB radio announcer. Hay claimed that he had received many phone calls from outraged citizens in support of his campaign against the HERO Parade in Queen Street. When questioned by another radio host, Paul Holmes, on Auckland's 1ZB radio station, he admitted to receiving only six calls (Holmes 1994).

Not only did queer bodies in the parade challenge hegemonic notions of heterosexuality, but also their very presence in the main street of Auckland was a challenge. How did the paraders act in Queen Street? The first HERO Parade did have a couple simulating sex in a large transparent balloon. There were men dressed in leather jock straps which showed their bare buttocks. Men had whips and chains on the Demon float which was also the safe sex float. Their banner read: 'Fuck Safe, Party Hard' (Legat 1994). These floats were in a minority. Other floats and parade entries included gays and lesbians in Greek attire, The HERO Marching Boys, a Priscilla Queen of the Desert bus, HIV positive groups, Gaily Normal and a float containing 'lesbian mothers'.

In 1996, at the HERO Parade workshop, I met one of the men from the Demon float and interviewed him about his involvement in the first HERO Parade in Queen Street. Robert said that the 'Clamp Club' had organised the Demon float. Robert and other members of the Clamp Club had not intended to cause such controversy. Their float represented their belief that s/m practices were part of 'everyday' and 'ordinary' life:

Robert: But we were just representing part of the community that was, you know, basically that way, and we'd been doing that [s/m] for ages and we just thought it was fine. It was the first parade and nobody knew what to expect, or what sort of reaction they would get from the public, anyway.
(Individual interview, 7 February 1996).

Robert was shocked by the reaction and subsequent national fame he encountered for being part of the Demon float.

Robert: So, we just went about what we normally did for, for the Clamp Club itself (yeah) and the shows we'd normally do. So we told HERO what we were going to do and then all of a sudden David Hay went *ape shit*.

>   Lynda: Fantastic.
>   Robert: And even, even worse than that, I mean, I was, I ended up on television and they put my image on television and stuff like that and I ended up in *Metro magazine* and I kind of thought, well this is a bit [makes a disgusted face]. And I actually saw David Hay there filming everything and I just thought, well this is interesting. Why is this going on?
>   Lynda: He's getting evidence?
>   Robert: I just couldn't figure it out. I just thought it was a sleazy old man going round with a video camera trying to tape films, you know, so he could go home and masturbate but, um it was um, David Hay just trying to get evidence and saying how disgusting and horrible we were.
>   (Individual interview, 7 February 1996, emphasis in original)

Robert could not understand the extraordinary reaction to the Demon float. He explained to me what they were wearing and what they were 'exposing'.

>   Robert: People were wearing leather G-strings and showing their breasts and stuff. There was no exposure, well yeah, there were bare butts, but there were no exposed genitalia. Why would people say that bare bottoms and bare breasts are not acceptable? They should see the parades overseas. There are a lot more um, sexual parades, especially the heterosexual ones.
>   Lynda: Like?
>   Robert: The Mardi Gras in New Orleans and the carnival in Rio.
>   (Individual interview, 7 February 1996)

According to Robert, it is not the type of behaviour that David Hay and his supporters objected to, rather it was the type of queer body that was objectionable. Sexualised displays and nudity in internationally famous parades are acceptable for 'the public' if they are within the realms of 'heterosexual performances'. Robert's point about international heterosexual parades might be difficult to sustain in Aotearoa/New Zealand. Much of the general public, city officials, and concerned Christian groups may not have had the opportunity to view highly (hetero)sexualised parades such as the Mardi Gras in New Orleans or the Rio Carnival in Rio de Janeiro. The kinds of parades that are acceptable in Aotearoa/New Zealand are the sporting heroes, Santa parades, Centennial parades, and seasonal parades such as the winter festival, blossom festival, or romance festival. These festival parades are marketed as 'family' events and are listed as tourist events on the Aotearoa/New Zealand Tourism Board's web site: Passport to New Zealand, New Zealand Arts and Culture Events on-line (New

Zealand Tourism Board 1998). What is often the central focus of these parades, however, is the nubile virginal 'festival queen' dressed as if for prenuptial sacrifice in the white wedding gown. The acceptability of this heteropatriarchal image is that sex can be imagined to be what happens *after* the parade.

Legat (1994) attempted to tease out the 'facts' of the HERO Parade. She lists that:

- It was adult behaviour
- It took place after ten at night when children and young people are not in Queen Street
- Anyone who thought they might be offended could choose to avoid it by not being in Queen Street
- You can't preach safe sex without talking about sex itself
- It was no worse than anything you'd see on television at the same time of night
- They do it at the Sydney Mardi Gras and the local authorities endorse it unreservedly
- It's an entertainment. It's meant to be slightly risqué, like the carnivals in Rio and New Orleans.

(Legat 1994: 94)

Such reflection was not part of the Mayor or Deputy Mayors' assessment of the parade. For two months after the night of the parade the dispute continued and moved far beyond a discussion of whether or not some of the parade floats (s/m and safe sex performance) were acceptable. The argument centred on the demand, by the Mayor and the Deputy Mayor of Auckland, that gays and lesbians 'behave in public according to certain standards' (Legat 1994: 89). Clearly, this desire (and fear) of the Mayor and Deputy Mayor of Auckland is centred on a public/private dichotomy. Inappropriate bodies and inappropriate bodily behaviour should not be allowed in public, especially not in the main street of Auckland.

A full Council meeting was called to address 'the public' objections over the HERO Parade. A demonstration was held outside the City Town Hall during the council meeting. People of Christian faith protested and held placards which asserted: 'Say no to nudity'. Children were carrying banners that said: 'We don't need another Hero' (Gearing 1997). During the Council meeting, Deputy Mayor, David Hay, provided the video he had personally made of the parade to substantiate his objections.
Legat (1994: 91) claims that:

> what was really debated at the April Council meeting was not behaviour at the Hero parade but homosexuality itself. The persistent questions on condoms, on anal sex and on whether or not homosexuality is a choice [as opposed to a biologically given] made that very clear.

Legat (1994) described several of the people and the concerns which they raised at the council meeting. The issues centred on the bodies of the gay men. She noted that:

> Julian Batchelor of Operation Jerusalem, a cross-denominational evangelist movement ... launched into his testimony: 'I was there! I saw a group of men in G-strings and bare buttocks looking as though they were inviting penile insertion. I saw men in a plastic dome wearing G-strings and oiled all over, writhing in a suggestive and provocative way'.
> (Legat 1994: 91)

Behaviour and gay bodies seemed to be inseparable for the protesters against the HERO Parade. 'None of this confrontation would be taking place if the Mayor, Deputy Mayor and certain Councillors did not object to homosexuality' argued a gay activist in response to the furore (Legat 1994: 95). Calls by Councillors to homosexuals to 'clean it up' for the next 1995 parade, were disingenuous. Deputy Mayor Hay continually returned to his contention that homosexuality is a choice and Julian Batchelor continued to hold placards that read: 'Homosexuality is a thing God hates' (Legat 1994, 96).

The Mayor of Auckland also took issue with the HERO Parade. He wrote a letter to the editor of the *Sunday Star Times* (3 April 1994: 3) outlining his stance:

> I am not prepared to personally encourage homosexuality or support homosexuality as a lifestyle as an individual or by the Auckland City Council from city rates. I did support the council resolution to write off $5000 from the costs of city services related to the Hero Festival in recognition of its stated intention to raise funds for AIDS prevention and related works. I do not endorse or accept some of the behaviour I viewed on a video of the parade and I am passing letters I am receiving which contain requests for action on to the police as the only organisation which can appropriately handle specific complaints.
> (Mills cited by Keith 1994: 3)

There were 40 complaints about the Parade made to Auckland City Police. Only two of the complainers had actually attended the parade. 'The other 38 had merely read about it, heard about it, or seen a bum or tit or two on the telly. The police will take no further action' (Keith 1994: 3).

Deputy Mayor Hay was interviewed again and claimed he had support for his opposition to the HERO Parade in Queen Street:

> I just got a letter this morning on my desk signed by a family to say, you know, both Thomas and Mark were separately passing through the city centre. They came across the parade and they were both offended by

nudity, simulated copulation, etc. on our main street. Now that's straight from a letter signed by a family and those people are upset, along with thousands of others, over this sort of thing in our street.

(Hill 1994)

In this instance, 'family' becomes synonymous with heteronormativity. 'Our' street is again identified as a heterosexual street.

Some 'mainstream' Aotearoa/New Zealand newspapers also reported positive comments regarding the 1994 HERO Parade. For example, one headline read 'Heroes on parade fill city with music' (*New Zealand Herald*, 21 February 1994: 10). The report noted that one spectator said 'she thought it 'brilliant' that homosexuals were showing themselves to the community' (*New Zealand Herald*, 21 February 1994: 10). Similarly, there were letters of support for the HERO Parade and criticisms of the Deputy Mayor, printed in newspapers. One person wrote:

It is predictable that Deputy Mayor David Hay is concerned at the flashing of flesh and suggestive gyrations at the Hero parade. I watched the parade and found little difference in the costuming (or lack of it) between the participants and a large number of the parade spectators. Does Mr Hay want to ban the parade or the citizens who inhabit Queen St. on a usual Saturday night?

(Burford, cited in *New Zealand Herald*, 2 March 1994: 2)

This letter highlights a blurring of the boundaries between paraders and tourists watching from the side of Queen Street. When the boundaries are dissolved between watchers and paraders, the foundation for complaint shifts. It is less likely that complaints will made by people who are participating in the entertainment.

Members of Auckland's gay and lesbian communities ran their own campaign against the Mayor and Deputy Mayor of Auckland. They distributed a flier which featured quotations from various media. The flier urged people to write to the Auckland City Council stating they supported the HERO Parade and that they objected to Hay's attempts to impose his personal views on Auckland City Council politics.

In 1995 the second, and somewhat sanitised, HERO Parade was held in Queen Street. The HERO project requested $10,000 funding from a city festival grant but the Council voted against this, and offered a waiver of the $5,000 service fee for the Council cost of cleaning of streets after the parade instead (Legat 1994). The HERO banner, which the Council agreed could be hung across Queen Street to promote the parade, appeared without any reference to safe sex, without the words lesbian or gay and without the Auckland City logo (Legat 1994). The Mayor of Auckland, Les Mills, was invited to take part in the 1995 HERO Parade. He declined. A supporter of the parade wrote a letter to the *North Shore Times Advertiser* stating:

The 1995 Hero Parade was judged by many (including the police) as being far from immoral, and the comment about the Mayor of Auckland declining to participate does not make him a better man, but proves his unsuitability to be mayor of such a fine city. Hopefully he will soon be replaced.

(Shaw cited in *North Shore Advertiser*, 10 August 1995: 7)

The HERO Parade, despite being very popular and attracting thousands of tourists, was an event that Auckland's 'city fathers' were still not proud of.

Abandoning the CBD

This opposition to the parade from Auckland's Mayor, Deputy Mayor and some Christian Councillors and Christian groups was, in part, behind the decision to shift the HERO Parade from the main 'straight' street of Auckland to 'gay' suburban Ponsonby. This decision prompted another public debate which was tracked by local newspapers. '"Keep the queens in Queen Street" is the rallying cry of a group of gays and lesbians who object to the announced change of route for next year's Hero Parade' (Sanders 1995: 3).

A petition was organised to retain Queen Street as the HERO Parade route. The petition co-ordinator expressed her displeasure at the plans to re-route the HERO Parade away from Queen Street to Ponsonby Road by stating, 'every major parade in the country heads down Queen Street and if it's good enough for the America's Cup, it's good enough for us!' (Cameron 1995: 3). The group believed Queen Street to be the 'bastion of heterosexuality' and having the parade in Ponsonby Road would 'detract from the symbolism and visibility of the parade' (Sanders 1995: 3).

Letters to the editor of *express* debated the decision to move the parade from Queen Street to Ponsonby Road at length. These letters could be seen to represent sectors of the gay community. In an *express* letter entitled 'Back to the closet', M. Stevens (1995: 2) favoured Queen Street over Ponsonby Road for the HERO Parade, saying:

> It was with a mixture of bewilderment and outrage that I learned of the decision to move the Hero parade from Queen Street to Ponsonby Road. Ponsonby Road is not the 'heart of gay Auckland' as some would claim. It has some gay-owned businesses and a gay bar on it. So what? I would say that for most gay Aucklanders, Ponsonby does not figure in their lives. If you are a middle-class white gay who can afford the high prices, then maybe, but that only describes a minority of us.

This *express* reader raises some interesting issues of class and 'race', as well as sexuality (and I suspect masculinity). He claims that Ponsonby has a gay element, but that it is an expensive suburb which caters only to those who can afford the entertainment (usually Pākehā, middle class, gay men). He continues:

> Even if Ponsonby Road were like Castro [San Francisco], the parade should remain on Queen Street. We deserve it, and should not bow to the pressure from the homophobic city council. Last year's [1995] parade was an outstanding success and helped make us a more visible part of the city, heads held high and proud. The logic of the parade organisers now appears to be 'back to the would-be ghetto and let's not offend anyone'. Back to the closet.
> (Stevens in *express*, 1995: 2).

Here the *express* reader identifies Ponsonby Road as a gay ghetto, or the (private) closet, and conversely Queen Street is constructed as a non-gay public street. Stevens continues with a further attack on the 1996 HERO Parade organisers:

> In my opinion, these people would have been handing out pink triangles with instructions on how to sew them on and then guiding us to the cattle cars if this were Nazi Germany. Anything to avoid a nasty scene. We deserve a huge riotous parade in the heart of the city, on Queen Street where every one else has their parades. Anything less is a victory for David Hay and his cohorts and a betrayal for all of us.
> (Stevens in *express*, 1995: 2)

Queen Street is again evoked as the most prominent place for parades of celebration and gay visibility. Another letter reminded *express* readers that the parade was a political and contested event:

> It is a political event. We do not have a Parade merely to show ourselves how wonderful we are – we know that already. We do not have a parade merely to publicise ourselves to ourselves – we know we are here. What we need to do is remind the wider community of our presence as well as to celebrate our presence ... What better statement is there than to have the HERO Parade down Queen Street, the main street of the city[?]
> (Christie 1995: 2)

Another criticism from people opposed to the Parade's move from Queen Street to Ponsonby Road reported in *express* was that 'Ponsonby Road was less well lit and less accessible to the general public, both in terms of transport and space for the general public' (Caldwell 1995: 3). Clearly, this letter to the editor of *express* wished the HERO Parade to go on attracting large numbers of the 'general public' (read: heteronormative).

The previous political struggles over the 1994 Queen Street parade between Auckland City Councillors and the HERO Parade supporters prompted further discussion by Dennis Brimble (1995: 2):

> Shock, outrage, anger; the superlative with which to describe how I feel over the HERO organisation's decision to move the 1996 parade to Ponsonby eludes me. The 'Community' fought long and hard writing letters attending meetings and a million other ways to ensure that we could 'claim our place in the sun'. Is all that effort going to merely be a 'moral' victory? Surely the adage 'Justice must not only be done, but seen to be done' must carry some weight.

Brimble (1995: 2) clearly links sexualised embodiment and Ponsonby Road when he remarks:

> I'm sure that the HERO Parade would be stunning as ever no matter where it was held but, to hold it on 'our own turf' so to speak would smack of masturbation and also only ghettoise the parade in that situation. What is the point of swapping one closet for another?

'Our own turf' is a reference to Ponsonby Road, recognised by many *express* readers as Auckland's gay(est) suburb. The reference to masturbation reinforces Ponsonby as the place for sexualised, private, homosexual acts.

If Ponsonby is 'our own turf' for homosexuals, then it may be considered as 'private' space. If this 'private' space is linked to a sexualised bodily act then it becomes a taboo private space – a space where only initiates might go – a ghettoised space. The string of connections, between 'body', 'private', and 'appropriate space' that appear in the media debates over 1996 HERO Parade site, confirm the positioning of HERO on the feminised side of the mind/body binary. The flipside of these binaries, the public masculinised place of the CBD is confirmed as the hegemonic, mindful site of 'ordinary' and 'everyday' heterosexuality.

The telephone poll, the media commentary and letters to newspapers indicate that the sites of gay pride parades for homosexualised bodies are politicised and contested.

I conducted a focus group interview with Rainbow Youth while they were painting their banner for their float in the 1996 HERO Parade, the first parade scheduled for Ponsonby Road. They felt their float would have more impact if the parade was to go down Queen Street again. Nick brought up the topic of Queen Street versus Ponsonby Road:

> *Nick:* The other thing is like, if we were going down Queen Street it [the float] would have a lot more impact.
> *Lynda:* Yeah, you think so?
> *Nick:* Yeah. I personally think the move to Ponsonby Rd is a negative one.

Figure 5.2 Collage of newspaper articles (source: *express* newspapers, September–November 1995).

Rachel: Yeah, Yeah
Lynda: Bit of a gay ghetto thing?
Frank: I think, yeah, it's becoming a bit inclusive when we have it in our own sort of district.
Rachel: Yeah.
Nick: Um, yeah and for us, for us it is not entirely a political move, I think any pride march at its very essence is a political move. But what's the point of saying *we're wonderful to ourselves*?
Lynda: Yes.
Nick: when we know we're wonderful and it's important to say we are wonderful to ourselves but, it is also important to say to the rest of the world that we are wonderful. That's what I think.
(Rainbow Youth Focus Group interview, 14 February 1996, emphasis in original)

Rainbow Youth's float did make an impression in the Ponsonby Road HERO Parade and I discuss this in detail in Chapter 4. In the next section I draw on data to explain the significance and impact of having the gay pride parade in Ponsonby Road. Ponsonby becomes constructed as home territory for gays and lesbians, thus the dichotomy of queer bodies in a straight (main) street is elided.

Suburban retreat

The HERO Project Director was not supportive of the first petition which advocated keeping the parade in Queen Street. He believed that: 'the petitioners are asking the community to forego the benefits of having the parade in Ponsonby Rd for the sake of a political statement' (Johnston cited in Caldwell 1995: 3). His stance may well have overlooked the core concern of the petitioners that the pride parade was believed to be, and designed to be, a political statement.

The petition to keep the HERO Parade in Queen Street was met with another petition to support holding the parade in Ponsonby Road. Arguments about the re-routing of the HERO Parade continued. Defending the new Ponsonby Road parade route, HERO Project Director argued that it would pass through Auckland's gayest suburbs. 'It's the area we party, socialise and live in ... This will increase our sense of ownership, pride and enjoyment of the parade' (Johnston cited in Caldwell 1995, 3).

Auckland City Council candidate and lesbian, Jo Crowley, was a supporter of the HERO Parade being held along Ponsonby Road. Crowley, standing in Western Bays ward which includes Ponsonby, argued that the lesbian and gay community has been treated 'with hostility and unfriendliness by the council, and we can stand around and be insulted by them or put money back into our community' (Crowley cited by Caldwell 1995: 3).

Crowley reinforced Ponsonby as a gay neighbourhood and Queen Street as a straight environment when she stated:

> The council has abdicated its responsibility to diversity and vitality in the city. The only things that get tickertape parades in this city are the things that Les Mills likes. Western Bays is the capital of the gay community, so let's go where we are welcome until the council can open its doors again.
>
> (Crowley cited by Caldwell 1995: 3)

It was also reported that an advantage of holding the parade on Ponsonby Road was that more money from HERO sponsorship, such as the alcohol company Dominion Brewery (DB), could be gained (Caldwell 1995: 3). The HERO Project Director stated: 'DB will give us more money as it works better for them in Ponsonby Rd – they will be able to sell alcohol along the road during the festival and after the parade' (Johnston cited by Caldwell 1995: 3). He also believes that holding the parade on:

> 'home territory' will support the businesses who support the community, and will add to the festive feel of the area during Hero. The biggest problem with the parade is that it's under-funded. This way we will get a better parade.
>
> (Johnston cited in Caldwell 1995: 3)

The metaphor of 'home' serves to connect gay bodies to gay neighbourhoods. When I interviewed the HERO Project Director about the parade route change, he talked of Ponsonby Road as a familiar friendly home.

Project Director: Primarily, I think the benefits of going Ponsonby Road is that it takes us back into our own homes, so to speak. It's a friendly atmosphere. It's an atmosphere and place that you know. A place that we live in and socialise in, all that sort of stuff.

Lynda: Sure, sure.

Project Director: It adds so much more to the festival for the whole monthly period. People can decorate their buildings more and it will give it a sense that the HERO Parade, the HERO festival has a heart, so to speak.

(Individual interview, 22 September 1995)

Queen Street is represented as the binary opposite to Ponsonby Road:

Project Director: Um, and the disadvantages for me for Queen Street are that it is an uninviting, cold environment and it is the people who wish to have it down Queen Street that are coming from, purely from the perspective of a political statement. So what they are saying is that they are prepared to forsake all those other sort of community things for all that political visibility.

(Individual interview, 22 September 1995)

'Cold' and 'uninviting' are attributes held in relation to the attributes of Ponsonby Road represented as the private 'friendly home' and the embodied 'heart' of the gay community.

The economic consequences of the HERO Parade travelling along Ponsonby Road, rather than Queen Street, are reiterated by the Project Director:

Project Director: There are financial advantages of going along Ponsonby Road. Our major alcohol sponsors will provide the rebate for outlets on Ponsonby Road ... The other side of it, probably the disadvantage of it, is that ... it won't make as bigger public statement as going down Queen Street. But it is not like either/or. It'll be, public statement going down Queen Street is that big [gestures wide with hands], and going down Ponsonby Road is that big [gestures smaller with hands]. It is still going to be a public statement. It is still going to be 100,000 maybe, a couple of hundred thousand people, so it will still be a statement, it will still get the proper coverage from the media.

(Individual interview, 22 September 1995)

Johnston (cited in Caldwell 1995: 3) 'hopes the route, past many of the city's gay venues and businesses, will be decorated with banners, posters and shop-front displays'. The 1996 site of the HERO Parade, Ponsonby Road, was marked with rainbow flags and rainbow bunting, HERO posters in shop fronts and billboards promoting safe sex along the roadside (see Figure 5.3).

One of my duties as a volunteer for the HERO Parade in 1996 was to distribute an 'Open Letter to all Ponsonby Businesses' written by the HERO Parade Director:

We are asking all businesses to support the parade in some way, however small (or large!) that support may be. Here are some suggestions of how your business can participate and support HERO:

1 Have a HERO window display
2 Display rainbow bunting for your cafe/store front
3 Offer a HERO day special to your clients.

This flier had an immediate effect. A floral business that I distributed this flier to wanted to sponsor the HERO Parade by offering spot prizes to the best

Sex in the suburbs or the CBD? 95

Figure 5.3 HERO Parade's safe sex billboard (source: *New Zealand Herald*, 9 February 1996: 2).

dressed visitor to Ponsonby Road on the night of the parade. Also, on the night of the parade, the floral business provided single flowers for sale with profits going to the New Zealand AIDS Foundation. These were the types of benefits that the HERO Project Director had hoped for when the decision was made to relocate the HERO Parade from Queen Street to Ponsonby Road.

Businesses did 'dress' their windows and shop fronts with HERO posters and rainbow colours. These visual indicators of the HERO Parade were also supported by various sound systems placed along Ponsonby Road as pre-parade entertainment. The letter to Ponsonby businesses stated: 'The parade will generate a carnival atmosphere in Ponsonby that night and we hope that you both enjoy and benefit from the extra patronage on 17th February 1996.' This promise, of a particular prestigious place-image, encourages businesses to 'site production within, and link their product with, a particular locality' (Edensor and Kothari 1994: 165). HERO sponsors, who produced alcohol products such as Heineken, Absolut Vodka, Deutz and Delegats, stocked Ponsonby Road bars with their products. The public was exhorted to:

> Purchase any of these brands: Absolut Vodka, Deutz, Delegats and Heineken at participating Ponsonby cafes, bars and restaurants during the HERO festival and a percentage of your purchase price will be donated to the HERO Parade.
>
> (*HERO Magazine* 1996: 3)

The Parade Director (quoted in *express* 7 December, 1995: 6) reported: 'We want to turn Ponsonby Rd into HERO Rd so that there's no way anybody who comes to Ponsonby Rd in the month before doesn't know that the parade is happening.' This is another way that the HERO Parade on Ponsonby Road becomes promoted as a tourist site.

When comparing the responses of businesses and residents of Ponsonby to the reaction of the HERO Parade in Queen Street, it became clear that a gay pride parade along Ponsonby Road became less of a political statement about gay rights and more about a night of entertainment. Queen Street was retrospectively constructed as a site of protest where queer bodies, out of place, had a political message to convey. Ponsonby Road became constructed as a site of performance where queer bodies were at 'home'. The tourist site of Ponsonby Road seemed to add further 'authenticity' for heterosexual tourists. They could now watch queer bodies in their 'natural' settings.

There remained, however, some controversy to the build up of the 1996 HERO Parade along Ponsonby Road. There were still some groups that resisted gay, lesbian, bisexual and transsexual bodies on parade in Ponsonby, despite Ponsonby being a well-known gay neighbourhood. A HERO Parade billboard, promoting safe sex, offended the Samoan church leaders of St John's Methodist Church (see Figure 5.3) when it was placed in the garden next to the church. The billboard featured the bare buttocks of a male body and the words 'No Ifs. No Butts. Use Condoms Everytime'. Reverend Paulo Ieriko told a *New Zealand Herald* (Ely 1996: 2) reporter:

> We live in a tolerant society and we are tolerant of other people's values and how they see their sexuality, but to impose it on others, that's where the church comes in ... I think it puts people off homosexuality instead of making them sympathetic to the cause.

The Ponsonby Plant Centre across the road from the billboard said some of their customers had found it offensive and one had suggested drawing a kilt on the naked body. Most believed, however, the 'billboard was worth it if it saved lives' (Ely 1996: 2). HERO Project Director said: '[w]hen promoting safe sex you have to talk about sex, just as when it's drink driving you have to talk about driving and drinking beer' (Johnston cited in Ely 1996, 2).

The billboard was damaged by members of the Methodist Church on the Sunday before the HERO Parade, 11 February 1996. The billboard underwent repairs and a new message was added. An acrostic, using the words 'embarrassment, homophobia, intolerance, bigotry, ignorance and discrimination' appeared in its place. Following these changes, Auckland City Council also demanded that the billboard be removed from its position beside the church as the owner of the property had not obtained permission for the billboard. The billboard was placed in the All Saints Church across the road from the Methodist Church. An Anglican minister gave his blessing for the billboard, saying 'It's [the billboard] still on about safe sex but it now confronts other, deeper issues in our society' (Whyte cited in Moore 1996: 1).

In this example, the Samoan church is a 'relic', in a sense, of the working class status of Ponsonby prior to the gay gentrification which displaced the poorer, many non-Pākehā, and Pacific Island communities, who lived in Ponsonby. There are still parts of Ponsonby that are 'privately' Pacific Island, rather than 'privately' gay. The Samoan church and billboard example cannot necessarily be entirely understood as predicated on public/private, straight/gay binaries, but as contingent upon historical, class, 'racial', and place politics.

In the last few days before the first Ponsonby Road parade the police issued a press release. The headline read "Keep it seemly' – Police' (*express*, 15 February 1996: 7). The article stated that caution was needed on the night of the 1996 HERO Parade:

> 'Keep the Hero parade seemly even if it's in your back yard' . . . Inspector Derek Davison says the new Ponsonby route will make the parade more colourful than the 'austere' Queen Street area. Ponsonby is a 'warm area for the gay community: they feel comfortable there. It's almost a traditional area'. But, he said, media coverage means people all over the country will watch the parade and that means, caution is needed. 'If there are opportunities for people to chuck rocks at it, they'll take them. My view is . . . let's make it an event which is a happy one, a careful one, but not to step over the bounds of community acceptability.'
>
> (*express* 15 February 1996: 7)

This newspaper article nicely sums up my assertions that the Ponsonby Road parade is considered, by controlling authorities, the correct place for gay bodies. The threat of television coverage, however, means that the parade will be in 'public' eyes and be judged by the 'community'. Hence gay bodies need to 'behave' and 'act appropriately'.

I have been arguing that relocating the HERO Parade to Ponsonby Road and away from Queen Street, restates the public/private dichotomy. A gay pride parade in Auckland's 'gay' suburb does not threaten or destabilise Auckland's 'straight' CBD. A gay pride parade further marks Ponsonby Road as other/different/exotic and more 'authentic' for tourists. The refusal to hold the parade in Queen Street reinforces the mind–straight/body–gay dichotomy.

In summary, the HERO Parade sites and debates surrounding the relocation of the parade from Queen Street to Ponsonby Road, foregrounds the problematic position of private bodies in public places. The connection between suburbs and gayness usually runs counter to the idea of suburbia belonging to heteronormative subjectivities (Hodge 1995). The newspaper articles and letters to newspaper editors clearly identify the specific ways western hierarchical dualisms are articulated and inscribed on bodies and places. The discourses that emerged from debate over the parade site can be structured into the following dichotomous relationships.

Queen Street – Ponsonby Road
Central Business District – Home
Uninviting – Friendly
Protest – Performance
Heterosexual – Gay
Business – Pleasure
Everyday – Exotic
Public – Private
Cold – Warm
Clean – Dirty
Mind – Body

The economic link between place and tourism was just one facet of the decision to relocate the HERO Parade from Queen Street to Ponsonby Road. Parade organisers believed that holding a parade in Auckland's gay suburb would create a warmer and more festive environment for the gay community. The Parade became more a form of entertainment than being about gay rights, or a celebration of gay pride. Queen Street became conceptually aligned with the 'mind' and Ponsonby Road with the 'body'. Moving the site of the Parade helped established the HERO Parade as a tourist event.

Conclusion

This chapter identifies that gay pride parades have the potential to 'queer' streets. Queering of streets creates a spectacle of queer bodies which many heterosexual tourists are 'drawn to'.

More substantively, I argue that the debate over the HERO Parade Queen Street site of 1994 provides evidence that queer bodies are often considered to be untrustworthy in the main (CBD) streets of Auckland. Bodies of gays, lesbians and transgenders are not considered by some to be 'appropriate' bodies to inhabit public streets. Queen Street, as a consequence, can be conceived of as representing the 'mind'. Streets of Auckland that can be conceived of as representing the 'body' are those streets in the gay neighbourhood of Ponsonby. Both hegemonic and transgressive notions of sexuality and sex, which are effects or results of political investment in the regulation of bodies, are at work at HERO Parade sites. In order to avoid the construction of a hegemonic, reified and static sense of HERO as a tourist site, it is necessary to recognise that localities are 'replete with internal differences and conflicts' (Massey 1993: 67).

By recognising and destabilising western dualistic thinking, I have highlighted the ways in which power struggles help to decentre static concepts of 'ordinary' and 'everyday' places. This account of the HERO Parade site works to consciously manipulate notions of gendered/sexed and sexualised embodiments in order to subvert hegemonic ways of 'seeing' sites without imposing a universal alternative.

6 Cities as sites of queer consumption

This chapter extends the book's borders to that of cities as international sites of queer tourism. In many western cities queer residential and commercial zones have become increasingly visible and attract a diverse public. Part of the visibility can be attributed to the success of gay rights movements and the economic recognition of the 'queer market'. Due to gay pride the changing politics of sexuality have meant that there is an increasing commercialisation and commodification of queer lifestyles. The growth of queer tourism and the greater visibility of lesbians and gay men as consumers have attracted critical attention (see Binnie 2004; Cieri 2003; Puar 2002a, 2002b, 2002c; Richardson 2005). The neoliberal rise of the cultural – queered – citizen is under consideration in this chapter.

Jasbir Puar (2002c) is discouraged by the celebratory tone of queer visibility politics that persists in queer tourism knowledges. She notes that from the literature queers are seemingly 'proud to be travelling and especially proud to be viable consumers in global, international travel' (Puar 2002c: 935). Some reasons are offered for this celebratory tone, in particular Puar (2002c: 935) argues that: 'discussions on gay and lesbian tourism are not yet contextualised in terms of neocolonialism'. This may be due, in part, to a focus on European and North American queer tourism spaces which are understood as positive disruptions of heterosexual space. Such studies fail to take into account the intersections of class, race and gender. 'While it is predictable that the claiming of queer space is lauded as the disruption of heterosexual space, rarely is that disruption interrogated also as a disruption of racialised, gendered, and classed spaces' (Puar 2002c: 936).

I am concerned with acknowledging and theorising difference in queer tourism spaces. In particular, I have tried to show in this book the ways in which pride parades may be considered emancipatory events *and* have the potential to reinscribe a number of hegemonies. To this end, I, like Puar (2002c) am both dismayed and encouraged with the effects of gay pride parades and their relationship with neocolonial impulses to travel. On the one hand, tourist spaces of parades provide an opportunity to queer – I would argue – heteronormative space. On the other hand, a celebratory approach to queer tourism and queer spatiality tends to occlude questions of gender (race, class and so on). Yet inclusion of these differences is crucial. These internal differences/tensions of queer are explored through an examination of cities as international sites for queer tourism.

The queer spaces of the cities that I refer to in this chapter – Auckland, Sydney, Edinburgh and Rome – may share some common characteristics but they are certainly not homogeneous. These spaces are linked through pride celebrations and queer iconography, but the differences are considerable. My discussion of international queer tourism sites aims to problematise further the supposed stability of gendered, raced, classed and sexualised bodies. I argue that the performativity of bodies are constructed by their spatial context (Bell *et al.* 1994; Brown 2000; Longhurst 2001a; Nelson 1999; Sedgwick 1990). Tourist sites, locales, cities and nations are unstable and come into being through socio-spatial enactments. Socio-spatial relations, therefore, do not simply differ between cities; rather, performances, spatial relations and interactions (re)produce places (Hubbard *et al.* 2002; Massey 1994). Within these unstable tourist places, bodies and space become gendered, sexualised, racialised, classed and so on.

The objectives for this chapter are as follows. First, under the heading Promoting Difference, I focus on particular tourist studies' literature that may provide an entry into understanding Auckland, Sydney, Edinburgh and Rome as queer tourism cities. I also draw on sexuality and urban space literature in order to highlight city politics of pride (Bell and Jayne 2004; Bell and Valentine 1995; Namaste 1992).

The contradictions that exist in Auckland and Sydney as queer tourist destinations are my first case study. Advertisements, popular press and interviews are used to elaborate on the promotion of these 'down-under' queer spaces. Auckland's material and discursively gentrified 'gay ghetto' – Ponsonby – has come to be, in part, associated with Pākehā, upper and middle-class gay masculinity. This sits awkwardly with the history of Ponsonby as a once impoverished space of poor ethnic enclaves.

Sydney's parade is held along a well-defined and visible gay (mostly male) commercial and residential district in and around Oxford Street (see Figure 6.1). This site may be read as already transgressive of the unarticulated 'everyday' heteronormativity in the streets of Sydney. The success of Mardi Gras and the incorporation of the queer celebration into city promotions and mainstream media may be understood as a kind of 'homonormativity', where once transgressive political displays are now corporatised, regulated and controlled. Debates on urban governance and sexual citizenship inform the construction of Sydney's queer space.

Another city with a famous international tourist reputation – Edinburgh (see www.edinburghfestivals.com) – is also examined in this chapter. Pride Scotland, however, remains marginal to city tourism promotion. Empirical material suggests that those who perform in Pride Scotland are ambivalent about the success and politics of the event. As a result of the lack of Pride promotion by Edinburgh city and Pride officials tourists are 'caught by

102 *Cities as sites of queer consumption*

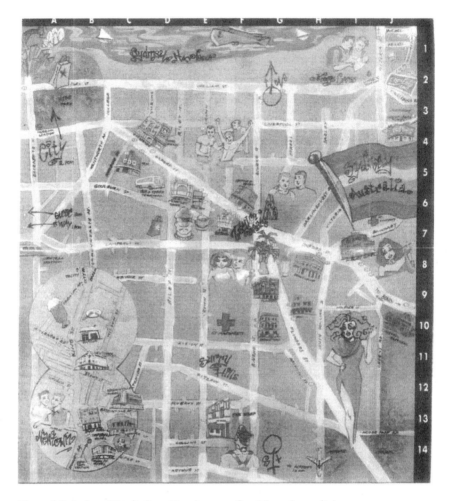

Figure 6.1 Sydney Mardi Gras Map (source: Gay Maps Australia).

surprise' and Pride performers are suspicious in what might be called an 'accidental tourist event'.

The final focus of the chapter is on the construction of Rome as an international queer tourism destination. Two 'world' tourism events were held concurrently during 2000, World Pride Roma and the Vatican's Jubilee (a millennial celebration claiming 2,000 years of Christianity) and I participated in these, as a tourist and a researcher. The celebration of queerness and Catholicism during the summer of 2000 throws into relief several conflicting discourses and establishes Rome as a tourist site of multi-layered and mobile performative geographies. I highlight the continuities and discontinuities between queerness and Catholicism.

Not all cities are equally enabling as queer tourism spaces and each contain internal differences that discursively and materially construct urban spaces. In all of these sometimes disparate, sometimes comparable city examples, I am mindful of the ways in which queer city destinations based on Pride celebrations, construct and maintain dominant discourses of queer subjectivities that may reproduce white, urban, middle-class landscapes.

Promoting difference

Until recently, tourism studies' academics rarely discussed the ways in which gender, sexuality and place are mutually constituted. Tim Edensor and Uma Kothari (1994) are an early exception. They studied the place packaging of heritage attractions at Stirling, Scotland. They focused on the Bannockburn Heritage Centre, the Wallace Monument, the Regimental Museum of the Argyll, and Sutherland Highlanders at Stirling Castle. They confirmed that these tourist sites: 'articulate masculinised notions of place and identity, and male dominated versions of the past which privilege white, male, heterosexual experience and activity' (Edensor and Kothari 1994: 65). Edensor and Kothari (1994) provide some insight into the particular material and representational practices and processes which influence the construction of the tourist sites. This applies as much to parades as to castles. For example, the advertising of the pride parades focuses on the sexualisation of place (see the explicitly sexualised map and guide of the activities and parade route for the Sydney Mardi Gras, Figure 6.1). Edensor and Kothari (1994: 165) argue that through the advertising of place-images and the generation of symbolic metaphors, representational practices continually '(re)inforce, (re)configure and (re)present local identities'. Edensor and Kothari (1994) are referring to the local subjectivities of the hosts and their representation in tourist site advertising.

In other words, both the representations and material 'reality' of gendered and sexualised bodies and place are central to the construction of cities as sexualised tourist sites. Similarly, Vincent Del Casino and Stephen Hanna (2000: 24) examine representations and identities in tourism map spaces and argue that: 'identities are defined and contested, and at times naturalised, through representational practices and individual performances'.

There are distinct places, and means of organising places, associated with tourist sites. 'Certain places and sites (with their landscapes, social practices, buildings, residents, symbols and meaning) achieve the status of tourist sights because of their physical, social, cultural – and commercial – attributes' (Britton 1991: 462). In a discussion on tourism, capital and place, Steve Britton (1991) works towards a critical geography of tourism. There are clear links between the physical, social, cultural and commercial attributes at gay pride parade tourist sites. His focus, however, does not discuss the materiality of classed bodies at such tourist sites. Nor does he pay attention to the political ways tourist sites become gendered and sexualised.

Binnie (1995, 1997, 2004) has been concerned with sexuality and tourism for the past decade. He discusses the creation and promotion of gay villages and questions how social scientists might 'make progress in theorising such districts without misrepresenting them or reproducing stereotypical notions about gay affluence' (Binnie 2004: 163). Binnie notes that gay villages have been primarily promoted via gay media – the internet, guidebooks, word of mouth – and I would add gay travel agencies.

The promotion of gay spaces, however, does not just happen within gay media. Dereka Rushbrook (2002) argues that queer space is popular with non-queers and may function as another form of ethnic diversity. Queer space is 'tentatively promoted by cities both as equivalent to other ethnic neighbourhoods and as an independent indicator of cosmopolitanism' (Rushbrook 2002: 183). Rushbrook (2002) turns to an article entitled 'The geography of cool' in *The Economist* (2002) which outlines the growing trend of gay clubs and neighbourhoods. The *New York Times* published an international club scene tour guide to cities such as Paris, Berlin, Madrid and Amsterdam (Cohen 2000). The article reminds its readers that these 'cool places' are frequented by gays and straights. For example, the Greenwich club in Berlin is a place where:

> cowhide adorns the padded walls and a certain animal intensity is definitely in the air as couples, heterosexual and homosexual, admire each other over some of the best martinis and whiskey sours in the city. This establishment, full of Asian-Germans and African-Germans, gives a real sense of new Berlin, a city whose population is an exotic mix.
> (Cohen 2000: 91)

The establishment's desire for diversity stands in for what might be understood as 'sophisticated allure' (Rushbrook 2002: 184) as non-white and/or queer bodies provide cultural capital for the assumed cosmopolitan traveller. In this article, Asian and African are offered as Other to whiteness, and homosexuality is offered as Other to heterosexual. This example highlights the consumption of queers and queer space by a broader but not necessarily queer public.

Discussions of the vitality and controversy of pride parades need to highlight the fact that such 'controversial' displays in public space attract heterosexual tourists. Some accounts of the Montreal Parade record the disruptive potential of the parade and implicitly endorse their role as spectacles, for example (Bell and Valentine 1995; Namaste 1992). Bell and Valentine (1995) provide an example of ambiguous and ironic relationships between gay pride parades and hegemonic culture, claiming it is possible to read the 1991 and 1992 Montreal Pride parades to 'illustrate a number of contemporary debates in sexual politics' (Bell and Valentine 1995: 20).

In 1991 Montreal's queer activists rejected the 'official' route that ran only through the city's gay district. A march through Montreal's gay neighbourhoods would have, it was argued, affirmed and empowered Montreal's sexual dissidents 'without being challenging or confrontational to the city's heteronormative culture' (Bell and Valentine 1995: 18). Queer activists from Queer Nation and ACT UP chose to march through downtown Montreal (both organisations are not currently active, however, see http://welcomehome.org/rainbow/lists.html).

The following year also attracted alternative actions of pride marches by queer activists. In 1992, the Montreal parade organising committee set rules and regulations for the queer bodies in the parade, stipulating that:

> there is to be no cross-dressing, no exposure of breasts or buttocks, no displays deemed too 'vulgar' or 'erotic' and no flags ... As if the outlawing of extravagant fashion weren't enough, it was suggested that the preferred attire of parade participants be blue jeans and a white T-shirt.
> (Namaste 1992: 8, cited in Bell and Valentine 1995: 14)

There were dramatic reactions to the organisers' rules and regulations. 'Across the city, sewing machines ran all night and stores ran out of sequins, fishnet and eye liner ... Montreal's sexual outlaws and perverts dressed to kill' (Bell and Valentine 1995: 14–15). There was also a mobilisation around Queer Nation's famous slogan 'If you're in clothes, you're in drag' (Bell and Valentine 1995: 15). As a consequence 'irreverent combinations of subjectivities proliferated, included fags posing as dykes, dykes dressed as clone fags, and bisexuals pretending to be fags pretending to be lipstick lesbians' (Namaste 1992: 9, cited in Bell and Valentine 1995: 15). Drag queens provided homoerotic entertainment for tourists, against the wishes of the organisers.

> This contradiction – that drag is simultaneously disavowed and permitted – is perhaps best understood in terms of its situation and context. That is, drag queens are permitted in certain spaces, among certain people, at certain times ... [The parade organisers] want their type of drag, in the spaces they designate ... Clearly, the activist focus on the idea that drag is everywhere threatens precisely the border, boundaries and limits [of acceptability].
> (Namaste 1992: 9, cited in Bell and Valentine 1995: 15)

The Montreal example provides opportunities to highlight the processes of power and control at work in Auckland's HERO Parade site, the Sydney Mardi Gras site, Pride Scotland Edinburgh and World Pride Rome. These parades all have similar elements of border, boundaries and limits which are, however, specific to each parade site.

The notion that queering the streets upsets heteronormative behaviour is both a convincing and a disturbing argument for situating the politics of gay pride parades for tourists. Activities that upset the 'everyday' or the 'ordinary' have the potential to become spectacles. People seek out the queerness – the extraordinariness – of the parades, to gaze at the queer bodies and to confirm the queer presence of the parade participants through the act of watching them. This production of paraders and tourists has the potential to upset the 'ordinary' and 'everyday' notions of public spaces. The focus on the 'ordinary' and 'everyday' has not been attended to in tourism studies.

Gay pride parades are not just spectacles, however; as I have argued in Chapter 5, they may reassert mind/body and heterosexual/homosexual dichotomies in both public and private spaces. Gay bodies on (public) parade are discursively and materially constructed as deviant, and hence, create the space for parades to become tourist events. I return to this irony later in the chapter.

In the next section I run two arguments that are inextricably linked. I argue that not only do parades queer cities, but also these parades are actively produced and consumed as tourist sites. The queerer the streets the more popular the tourist attraction for gays and lesbians, but also, and more significantly, for heterosexual tourists. Cities which have socially exclusive governance policies – those which actively promote Pride events – turn queer spaces into regulated spaces that become subject to planning controls and hence state intervention (Binnie 2004; Richardson 2005).

Auckland and Sydney: down-under difference?

Subjectivities and parade sites are mutually constituted through the embodied representations and material 'realities' of the parade. The economic significance of tourism has a close association with particular representations of Auckland and Sydney. Economic practices include the investment of capital into the tourist site and infrastructure of Ponsonby Road in Auckland. Alcohol companies, restaurants and retail businesses target both the HERO Parade and the Sydney Mardi Gras sites and help reconstruct these places through their advertising. Some alcohol companies build their campaigns around political issues of gay and lesbian subjectivities. For example, two 'beverage' billboards feature photographs of an 'everyday' looking baby girl and boy accompanied by the slogans: 'Filthy Faggot: Your son might be gay. Don't discriminate' and 'Dirty Dyke: Your daughter may be gay. Don't discriminate'. A complex interweaving of politics, bodies, 'everyday' representations, consumption and economics establishes the parade routes as tourist sites.

I asked the Director of the HERO Project to explain the differences between the HERO Parade and other gay pride parades. First, he responded that: 'HERO came out of HIV promotion and health programmes' (individual interview, 22 September 1995). I replied:

Lynda:	Right. I have been surfing through the internet looking up all the pride parades in the world and a lot of them do focus on protest to mark the Stonewall riots, and it seems to give them a very specific focus.
Project Director:	It does. I mean London Pride is all about walking on the streets and blowing whistles, it's a, it's just a protest, protest slash pride statement. And I think the emphasis is more on protest than on pride. I went on the Washington [DC, U.S.A.] march and that was just about protest and pride. There is very little about performance when Mardi Gras and HERO parades are a lot about performance.... Being out there, if you want to be out there, or whatever, uh, and there is very little, actually there is very little about protest in our parade.
	(Individual interview, 22 September 1995)

Ideas around performance, HIV health promotion and entertainment are favoured by the HERO project rather than ideas about a march to protest for gay rights and arguably this is true for the Sydney Mardi Gras. The constructions of the HERO Parade and the Sydney Mardi Gras are culturally distinct from parades held in the northern hemisphere. The most obvious difference is the attention and regulation of parade entries to ensure that the parade is a tourist-centred event and that tourists will have a spectacular night of entertainment.

The focus of the HERO Parade and (as I discuss in more depth in the next section – the Sydney Mardi Gras Parade) is on performance and the audience is dominated by heterosexual tourist spectators. The HERO Parade has become an extremely popular event and attracts many tourists. HERO, according to the HERO Project Director, has always been concerned with performance.

Project Director:	It's about saying 'come along, this is who we are', in a *performance* sense, there is a little bit of a trick in that, a lot of people will think (//)
Lynda:	[Is it] real?
Project Director:	Yeah, people in parades don't look like that during the day (Yeah). Um, but it is performance for them and we enjoy doing it. Like, I was in the parade, in the Marching Boys, and it's just really great fun. The fun was like all the rehearsals with my gay friends. And then going out and performing in front of nearly, about 100,000 people. It's a performance.
	(Individual interview, 22 September 1995, emphasis in original)

Another indicator of the construction of HERO for the 'straight' public is that in 1997 and 1998 the full parade was presented in 'prime-time' on national television. The HERO Parade as public 'product' is now sold to television production companies and can be purchased as a video cassette. The Sydney Mardi Gras is also televised in Australia and marketed in video cassette form. Such products are advertised as tourist souvenirs in Sydney, along with T-shirts, tea towels, key rings and so forth.

There are a number of international (and some domestic) gay tourists who come to Auckland for the HERO Parade and associated activities such as the HERO Party and Festival. The Director of the HERO Project 'sells' HERO on an international scale:

> *Project Director:* I'm definitely selling the project. But I'm not altering the project to accommodate any product for tourism.
> *Lynda:* Right.
> *Project Director:* You know, but the product, mainly the festival and parade and party, and I'm going out and selling that. And we are doing quite a bit of that. I'm working with a company called 'Men on Vacation' which is the biggest wholesaler in California. It's gay men travel ... they wholesale about 500 people to Mardi Gras. But we are working with them. And I have managed to get a full page in the *Out* magazine, which is a commercial ... and it will have information on HERO. So we are promoting it in a sense. And HERO is tied in two weeks always before Mardi Gras, so they are combined.
> (Individual interview, 22 September 1995)

The international tour company, Men on Vacation, advertises tours from the United States to New Zealand and Australia which take in both the HERO and the Sydney Mardi Gras celebrations. For example, their web site (http://www.electriciti.com/tvlexprs/pages/menvac.htm) offers:

> New Zealand pre-Mardi Gras Tour: February 15th to 27th – The Heroes Party and The North and South Island. Join Men On Vacation for our best pre-Mardi Gras tour ever! We'll arrive in Auckland, the city of sails, just in time for The Heroes Party, Mardi Gras New Zealand style! After a day to relax, we'll explore the North Island from the Coromandel Peninsula to Rotorua. We'll then fly to Queenstown, where many adventures await, and travel up the west coast and over Arthur's pass to Christchurch. Plan now on joining Men On Vacation.

Men on Vacation is one of many travel agencies that specifically serve gay travellers. Another company that offers tours to Auckland's HERO and to

Cities as sites of queer consumption 109

the Sydney Mardi Gras celebrations is Above and Beyond Tours. They advertise The Wonder Downunder Tour (see Figure 6.2). In this advertising, too, the Sydney Mardi Gras Parade is constructed and represented as a tourist event and marketed as part of a tourist package product. Markwell (2002: 83) notes that: 'the intersection of the local and the global, and the

Figure 6.2 Wonder Downunder Tours (source: Tourism Australia).

tensions that flow from this intersection, can clearly be seen in the Sydney Gay and Lesbian Mardi Gras'. Mardi Gras, while retaining some Sydney uniqueness is bounded by a global gay and lesbian tourism industry that insists on consumption of experience, spectacle and requires considerable corporate sponsorship.

> Australia is a land of astonishing scenic beauty and intriguing cultural diversity. Here gay and lesbian communities are vibrant, creative, eclectic. But don't just take our word for it! We invite you to come visit us; to discover the wonder, the pleasures, the sights and the sensations yourself ... Every February, Australia's mid-summer, the effervescent mood of Sydney's gay community erupts into a joyous, month-long celebration – the legendary Gay and Lesbian Mardi Gras. Billed as the largest gay and lesbian festival in the world, the Mardi Gras is a month of culture and fun. Its climax is a dazzling evening parade through the streets of Sydney.
> (Above and Beyond Tours, 1996)

This type of tourist site representation articulates gendered and sexualised notions of place and subjectivity (Edensor and Kothari 1994). It also has the potential to guide sex tourists from around the globe (Binnie 1995).

An Auckland travel agent, who caters to the gay tourist market, explained that:

> There are gay people who spend their life partying and they go to HERO and then they go to Mardi Gras, and then they might go up to something in the States. And that's quite normal to do that, and just party all year. You know, go away and basically go to parties on holiday, but the party is the attraction.
> (Individual interview, 24 January 1996)

This tourist operator confirms that the majority of international tourists to HERO and the Sydney Mardi Gras would be gay. The majority of domestic tourists, however, are straight, as also explained by the tourist operator:

> People will say to you that *'no one can put a party on like a poofter.'* So, you know, straight people *love* gay parties. They, they love HERO. And they love Mardi Gras. And *generally* most people I would send to Mardi Gras would be gay ... But there are straight people that go over with their gay friends and have a good time over there. A lot of people go over to Mardi Gras to the parade, and the excitement, and the events and the shows, and theatre and everything else associated with it. Um, there are certainly party animals in the straight community who are very attracted to gay parties because they're always well done.
> (Individual interview, 24 January 1996, emphasis in original)

Gay pride parades and parties – therefore – have a reputation of 'being well done' or, in other words, well organised, well performed and risqué. Such a reputation reinforces and spectacularises HERO and Sydney Mardi Gras parades. The Sydney Mardi Gras Parade has a reputation as being the most popular gay and lesbian event in the world (Carbery 1995). Its popularity rests on its representation as a highly sexualised tourist site. One travel advertisement read: 'A pilgrimage to Mecca – Sydney and Sleaze' (*express*, 26 October 1995: 17).

Sydney has become successful in managing the production of 'celebratory spectacles, now as often for non-gay-identified consumption as gay-identified consumption' (Knopp 1998: 106). Knopp (1998) is referring to the popularity of the Sydney Mardi Gras as a tourist site for heterosexuals. Sydney also has well-defined and visibly 'gay' areas such as Oxford Street and the neighbouring suburbs of Darlinghurst, Paddington, Surry Hills, Glebe and Newtown (Leese 1993; Witherspoon 1991, cited in Knopp 1998). These gay neighbourhoods are mapped for tourists attending the festival events (see Figure 6.1).

Knopp (1998: 107) discusses gay male identity politics in Sydney and observes that:

> in terms of urbanisation, then, Sydney features an articulation with this particular form of identity politics that is surprisingly democratic. Only in a few other places might one find such a highly sexualised and politically successful gay male culture in an urban environment – perhaps Amsterdam, Copenhagen, or San Francisco.

The success of gay male culture, and 'Other' sexualised subjectivities in Sydney was evident when I visited in 1996. Many gay, lesbian and transgender events are organised in the month that precedes the parade and party. Drag queen performances such as 'Frocks on the Rocks' were held, Oxford Street businesses participated in an event called 'Shop yourself stupid', there were queer cabaret, films, bands and choirs, comedian acts, art shows, Mardi Gras costumes displays in Sydney's Power House Museum, and sporting events, all of which act to queer significant parts of Sydney during Mardi Gras month. Oxford Street, and its neighbouring streets, is decorated in Mardi Gras colours and images. Sydney has also hosted the Gay Games (Waitt 2003).

Bars, restaurants, bookshops, gay travel agents, beauty salons, legal and insurance services, fashion stores, music shops and so on are part of a number of businesses that specifically target queer Sydney customers. Such commercial commitment relies on, and is productive of, both a queer consumer-citizen and the development of an expanding gay business community. The extent to which the Gay and Lesbian Mardi Gras has become institutionalised and normalised, that is, almost part of the 'ordinary' and 'everyday', became obvious when a person passed me in Oxford Street and said to me

'Happy Mardi Gras'. Mardi Gras has become a sanctioned celebration with its own 'salutation'. One person responded to my Sydney questionnaire: 'Mardi Gras is for queers what Christmas is for heterosexuals.'

While the Sydney Mardi Gras festival, and in particular, the parade, acts to queer the streets of Sydney, it also acts to reaffirm homosexual bodies as Other to, and deviant from, the heterosexual norm. This happens in two ways. First, because the site of the parade is in 'gay' neighbourhoods, the parade (re)establishes the constitutive relationship between gay bodies in gay neighbourhoods. This is similar to the HERO Parade being sited in Ponsonby Road. The parade site is in an (already) sexualised suburb. Second, unlike gay pride parades that travel through 'main' streets of cities (Montreal, Pride Scotland, and the Queen Street HERO parades), the Sydney Mardi Gras Parade site does not challenge the 'ordinary' and 'everyday' heterocentricity of 'straight' streets. The Mardi Gras has become an entertaining and celebratory tourist event, rather than a purely protest event, which challenges heteronormative streets. The parade was born out of political protest on 24 June 1978 (Carbery 1995). Two gay Sydney activists had received a letter from the San Francisco Gay Freedom Committee urging gay communities around the world to organise events in the last week of June to coincide with the 1969 Stonewall riots of New York City. Many people were arrested at the first Sydney International Gay Solidarity Day celebrations and protests were held in the weeks that followed the first march (Carbery 1995). It was not until 1981 that the Parade date moved to summer (March) and the Sydney Mardi Gras began to build its celebrations around known gay streets and neighbourhoods (Carbery 1995).

Another geographer concerned with the construction of gay (male) space in Sydney is Kim Seebohm (1993). He discusses the underlying themes and meanings of the Sydney Mardi Gras Parade. Seebohm (1993: 194) suggests the 'paradescape' is spatially, politically and symbolically inscribed on the landscape. Seebohm (1993) indicates, however, that the hegemonic culture exerts a high level of social control over the gay community and Mardi Gras. Mardi Gras, it could be argued, is also resilient and resistant 'to the web of signification spun by dominant elites' (Ley and Olds 1988: 191). Seebohm (1993) also uses Eco's (1986: 6) carnival concept of 'authorised transgression' to claim that gay public rituals reinforce the political and social mechanisms of control that hegemonic culture exerts on minority groups.

> As a significant spatial event, Mardi Gras has played the major role in the spatial construction of gay space in inner-city Sydney, creating a territorial and symbolic centre for the gay community in Sydney. This construction should not be viewed as the gay community concentrating itself into inner-city space for its own protection. Alternatively, it acts as a statement of popular culture acknowledging and tolerating the gay community in spatial terms.
>
> (Seebohm 1993: 205)

I tend to agree with Seebohm (1993) that there are hegemonic controls over such a popular event. I want to add that the construction of gay space in Sydney has helped to attract tourists. The HERO Parade in Ponsonby became a more popular event for tourists as it was constructed and held in an 'authentic', exotic, gay location. The site of the Sydney Mardi Gras is also in gay neighbourhoods. This spectacularises the event and provides an environment which invites multiple and competing gazes.

The corporate links that are made in Sydney and Auckland can be seen operating in a number of USA cities. During gay pride week in Seattle, Starbucks' employees wear specially made pride T-shirts to represent the coffee company (Dahir 2001). Queer communities are considered a:

> $500 billion market, and it's growing fast. Gay shows are drawing advertisers on TV. IBM, Jaguar, Ford, Subaru, Sears, P&G, J&J, Kodak, Verizon, Wells Fargo, and United Airlines are buying space in gay and lesbian magazines. Motorola, GM, Merck, Neiman Marcus, and Virgin Megastore are popping up on websites the LGBT community visits. Coke, Delta Air Lines, and BellSouth strut their stuff at Atlanta's gay pride festival. And Bud Light has put its brand on gay rodeos.
> (Armour 2003: 60)

According to Steve Kates and Russell Belk (2001: 392) gay pride celebrations 'may possess a decidedly syncretic nature and may be usefully viewed as public carnivalesque festivals, culturally shared rites of passage, forms of politically motivated consumption-related resistance, and magnets for commercialisation within the context of the festival'. They offer a range of contradictions surrounding consumption and queer politics and argue that:

> conspicuous consumption during Lesbian and Gay Pride Day may be politically dubious activity, but it is difficult to criticise it unambiguously when consumers explicitly consider that this same display and show of market power may actually result in social legitimisation of gay and lesbian community.
> (Kates and Belk 2001: 415)

Kates and Belk (2001) hold in tension the ambiguities and complexities of consumption as a form of resistance to mainstream commercialism. I want to focus briefly on one of their conclusions, the idea of social legitimisation. The idea of homosexuality as culturally subversive is a popularly held notion. Queer bodies have long been thought of as threats to social stability and public order. As Binnie (1995: 189) points out legislation in Britain has 'served an important ideological purpose as part of the wider Conservative family value agenda in which homosexuality was seen as one root of moral decay of the nation'. This legislation, however, has been significant in the mobilisation of gay activists and allies, assuring sexualities remain a crucial focus of debate and struggle in particular nations like the UK (Bell and Binnie 2002).

Homosexuals have been constructed as in need of external control and regulation and hence, in this book, I have documented some of the outrage that has accompanied Pride. The recent emergence, however, of a new category of citizenship in public discourse – 'the normal gay citizen' – is worth considering. Diane Richardson (2005) highlights the shift in political agendas of social movements concerned with gender and sexuality. She states that rather than critiquing social institutions and practices that have historically excluded queers, the dominant discourse of the last decade has been that of seeking access to mainstream culture through demanding equal rights (Richardson 2005). Integration and assimilation strategies have been employed by groups to achieve social change. Gay and lesbian struggles over, for example, the civil recognition of domestic partnership, the right to marry, the right to serve in the military are at the forefront of debates within lesbian and gay groups (Phelan 2001; Richardson 2000). At the heart of these struggles is a desire to be seen as 'normal' and 'ordinary' and 'good' gay citizens. Parades and queer celebrations may be understood as enabling new forms of neoliberal cultural citizenship, where queers may want to buy into it. The commercialisation and commodification of gay and, to a lesser extent, lesbian communities over the last ten to fifteen years has fostered the emergence of new commercial and professional (gay) identities (Richardson 2005).

I will use an example from another Parade site, Edinburgh, in order to continue to work through these notions of sexual cultural citizenship, tourism and Pride celebrations. Edinburgh provides an opportunity to examine implications where resistance exists towards the neoliberal rise of the sexualualised culture citizen.

Edinburgh

The city of Edinburgh has a reputation for being socially conservative and, as Knopp (1998: 159) observes: 'has produced strong resistances to gay male visibility and power'. Edinburgh tends to be recognised as a more middle-class and conservative city than, for example, Glasgow. Edinburgh has a strong and symbolic Scottish identity being the centre for the Scottish Parliament and the Scottish legal system. Historically, city policies on urban development have favoured Edinburgh's middle-class through the preservation of the city's core (the Old Town) and surrounding neighbourhoods. Knopp (1998) notes that council housing and working-class neighbourhoods have been pushed to peripheral areas. Furthermore, the 1980s saw financial deregulation and administration reform that resulted in an upsurge of large quantities of capital being 'invested in highly speculative, short-term real estate projects in the city' (Knopp 1998: 160).

It is within this context that a particular conflict occurred in the late 1980s and early 1990s involving a sex scandal which focused on the Scottish judiciary.

> The local tabloid and mainstream press, in cooperation with certain local police officers, circulated reports alleging that the legal profession in Scotland (including the Scottish judiciary) had been infiltrated at its highest levels by a 'magic circle' of gay lawyers and judges who were conspiring to 'pervert justice'.
> (Knopp 1998: 160)

Sex crimes, mortgage and housing allowance fraud were some of the alleged offences. Knopp (1998) notes that despite many years of investigation the allegations were found to be baseless, but a particular 'scandalous discourse' remained that threatened the established Scottish identity of Edinburgh and the sanctity of the legal system.

There are several points to be made from this example. It highlights the powerful ways in which discourse constructs material places. Throughout the scandal press invoked metaphors of movement and penetration and specific sites were identified as to where 'sex crimes' allegedly took place. Knopp (1998) notes that gay bars and clubs; hostels, bed and breakfast lodgings; and cruising grounds such as the road in front of the Scottish Office, became publicised geographies of sexual subversion in Edinburgh. 'So in this case the discursive construction of gay spaces and places was much more straightforwardly a strategy of oppression than of resistance' (Knopp 1998: 160).

The scandal highlights some dominant discourses which construct the sexualised spaces of Edinburgh. Liz Bondi (1998) draws on an example of gentrification in Leith – a part of Edinburgh known for and by sex workers – to comment on the way in which the local press establish a distinction between sex workers and non-sex workers. Bondi (1998: 191) notes women who are not sex workers are presented as 'mothers, as wives, as workers, as bingo payers and so on, but never as overtly sexual beings'. I draw on this example because I think it adds resonance to the ways in which Edinburgh may be understood as conservative, middle-class, urban space. Women are not deemed to be sexual beings, according to the local Edinburgh press. Imagine then, reactions to the overtly sexed and sexualised bodies of Pride Scotland.

A concern about Edinburgh's conservatism was raised at one of the focus groups I facilitated. Twelve women (some identified as lesbian, some as heterosexual, some as bisexual, some did not specify their identity) compare and contrast their feelings regarding participating in Pride Scotland in Glasgow and Edinburgh.

Moira: I'm not sure, I mean they're entirely different cities. I mean Edinburgh's got so many tourists, especially on Princes Street, (hmm – group agrees) there's not a local community that's there. Whereas Glasgow is much more of a local feel around the area where we would march (mmm – group agrees).

Carolyn: It's about difference, I mean, someone was telling me that they were at a meeting in Glasgow, and they have taken the most incredible amount of refugees in Glasgow. In Edinburgh they're not even, they're quibbling about whether they're gonna take 100 or 150 refugees into Edinburgh. I just think Glasgow tolerates difference, and I think that Edinburgh is a really reactionary, conservative city.

(First Edinburgh Focus Group, 2 July 2001)

Edinburgh, well known as an international tourist city, is understood as intolerant to difference. Here difference stands in for queer bodies and non-white, non-Scottish bodies. Carolyn's comment regarding Edinburgh's resistance to accept refugees illustrates which bodies might be Othered in Edinburgh.

The women's drumming group commented on the 'accidental' nature of Pride Scotland (see Chapter 3) due to the lack of city sanctioned promotion and a systematic advertisement programme by Pride officials. Another participant, Kate, highlighted the difference between Pride Scotland and a parade she attended in Perth, Western Australia.

Kate: It was very different to here [Edinburgh] actually ... It's kind of a different set up than here, I think, over there [Perth]. It's much more like a carnival week sort of thing, it's not just a one off march. So we have emm, I think we have it on a Friday night which on the hill by the entertainment area, emm and it's just huge because it's more of a carnival parade than a marching parade, if that makes any sense?

Lynda: Yes.

Kate: You have all the people on bikes and all the stuff just going along. Emm, so that's huge, I mean there's more people watching it, and emm, it's much more people cheering and clapping, it's not just standing and watching. And people are also drunk watching, and you know, it's more of a party.

(Second Edinburgh Focus Group, 9 July 2001)

When this participant reflected on these differences between Edinburgh and Perth, she noted that the Perth Parade ended in a park where there were stalls and fund raising auctions. She also noted that Perth Pride had motorised floats and a bigger emphasis on display and entertainment. Perth's Pride celebrations appear to be integrated and accepted by mainstream culture. In Edinburgh Pride is marginalised, the Parade is small, and the event is not commercialised nor actively marketed to tourists.

While more and more western cities are becoming sexualised sites of queer consumption, Edinburgh does not seem to follow this trend. Remaining on the outside of city promotions creates some social acceptance anxieties for gays and lesbians. Rather than plea to be more integrated into Edinburgh, however, the participants critique the social institutions and governing bodies of Edinburgh, stating that Edinburgh is conservative and intolerant to difference. This small parade, Pride Scotland, confuses tourists but may be understood as a highly successful subversion of new forms of sexualized neoliberal citizenship.

My last example of cities as sites of queer consumption is Rome. Queer tourism sits uncomfortably beside Christian millennial celebrations during the year 2000.

Rome 2000: proud, queer and Christian?

World Pride was held in Rome in July of 2000, during the Vatican's Jubilee, or *Giubileo* – a millennial celebration claiming two thousand years of Christianity. These two events during Rome's peak summer tourist season meant that millions of pilgrims (Christian and/or queer) descended on the city. Rome was chosen for World Pride because organisers hoped to bring about equal rights for Roman and Italian gays and lesbians, as well as stimulate increased queer tourism (Luongo 2002). Holding World Pride in Rome during the Jubilee, organisers also hoped to establish dialogues with the Catholic Church and encourage it to change its anti-queer stance 'thereby affecting the gay and lesbian community throughout the world' (Luongo 2002: 167).

It is difficult to gauge the extent to which World Pride changed and/or reinforced Rome's anti-queer stance. Dialogues were established between the Catholic Church and gay and lesbian organisers of World Pride. While the mayor of Rome defended his decision to host the event, the Vatican disapproved of the city holding the event during the Jubilee but indicated that it would not intervene formally with the government to stop the celebration (Becker 2000).

Historically, Michael Luongo (2002) notes, the Mediterranean region attracted gay travellers from northern Europe and he draws on research (Aldrich 1993) to show that nineteenth century art revealed a fascination with young Mediterranean men in ancestral dress. Indeed, many European nations thought Catholicism was synonymous with homosexuality. 'Dutch policy makers of the early 1700s regarded sodomy as an imported 'Catholic, or more particularly, Italian vice' (Van der Meer, 1997: 2, cited in Luongo 2002: 168). Not surprisingly, many artists of the Italian Renaissance had same-sex lovers that were hidden or closeted. 'Michelangelo's family, for instance, tried to obscure his authorship of his love poems to men' (Cady 1992).

While other cities may be experiencing a rise in queer neighbourhoods and the neoliberalisation of queer consumption, Rome has not. Gay guides of Italy discursively produce homosexuality as secretive, invisible, unacceptable and Italy as conservative (Luongo 2002). The stance of the Vatican appears to be steeped in fundamental and conservative theology which follows a very restrictive interpretation of homosexuality and is derived from the supposed sacredness that links sexual acts to procreation. The Catholic Church's viewpoint can be found in the 'Letter to the Bishops of the Catholic Church on the Pastoral Care of Homosexual Persons' (McNeill 2003: 548–549):

> Although the particular inclination of the homosexual person is not a sin, it is a more or less strong tendency ordered toward an intrinsic moral evil; and thus the inclination itself must be seen as an objective disorder. Therefore special concern and pastoral attention should be directed toward those who have this condition, lest they be led to believe that the living out of this orientation in homosexual activity is a morally acceptable option. It is not.
>
> (In Fox 1995: 148–149)

The response from Catholic gays and lesbians ranged from sadness to amusement and it has been noted that the Vatican, rather than the Catholic Church *per se*, is at odds with changing social norms and moralities (Fox 1995).

Donald McNeill (2003) sifts through the politics of Rome as a global city during *Giubileo* and notes that this spectacular event cannot be divorced from the ongoing power relations between church, state and, I would add, the sexual citizen. The role of Rome's mayor – Francesco Rutelli – was crucial to the 'acceptance' and promotion of World Pride Roma (McNeill 2003). Rutelli was known for his acceptance of gay rights. For example, in 1994 he supported the Italian Gay Pride march containing approximately 70,000 people. It was reported that some marchers chanted '*Siamo tanti, siamo belli, siamo tutti con Rutelli*' (cited in Bruto (1997: 129) and translated by McNeill (2003: 549) as 'We are many, we are beautiful, we are all with Rutelli'). He had also appointed to his council a representative with specific attention to gay civil rights. Rutelli's personal conversion to Catholicism in 2000 coincided with his removal of overt support to gay rights (McNeill 2003).

In the lead up to World Pride Roma various groups made public their opposition to the parade. The march was deemed 'inopportune in the Holy Year' by the social democrat government leader (NcNeill 2003: 549). The opposition right wing alliance leader – Silvio Berlusconi – considered the event 'mistaken both for the timing and backdrop chosen' (NcNeill 2003:

550). Cardinal Angelo Sodano, the Vatican Secretary of State, said he was sure that Rome city officials would reconsider their decision to allow the march. 'There are certain rules of coexistence and good sense. Rome has symbolic importance as a holy city, and what happens here has repercussions around the world' (Becker 2000: 9).

The Pope's reaction came on the day after the march:

> In the name of the Church of Rome I cannot fail to express sorrow for the affront caused to the Great Jubilee of the year 2000 and for the offense against Christian values of a City that is so dear to the hearts of all the Catholics of the world.
> (La Rocca 2000: 2, cited in McNeill 2003: 550)

This is an explicit attempt to distance non-heterosexual sexualities from the celebration of *Giubileo*, and crucially, maintain the space of Rome as a sacred and global city. My experience of World Pride Roma was an outcome of the intersectionality of queerness and Christian-ness. In some ways the parade was similar to other northern hemisphere parades.

There were colourful costumes and gorgeous drag queens. People laughed, sang and obviously enjoyed the march. At the end of the march, at the Circus Maximus, there was a collection of tents erected temporarily to house some pride merchandise. The parade was deemed successful with the crowd numbers estimated at 200,000. There was, however, a distinct lack of religious parody – Pope and Nun costumes were absent from World Pride Roma – that would be commonplace in, for example, the Sydney Gay and Lesbian Mardi Gras Parade. I did find one parader who insisted that 'God is Gay' (see Figure 6.3). While the Vatican feared that the parade would turn into a provocation, organisers of World Pride Roma were sensitive to the Catholic Church and planned the parade to pass alongside pre-Christian pagan monuments, not churches (Poggioli 2000).

While many western cities are experiencing and promoting a type of neoliberal sexualised citizenship (and the associated spaces) Rome complicates the debate surrounding sexuality, tourism and Pride celebrations. In 2000 Rome 'the sacred city' became an exhibition space for a commercially and spiritually motivated Church. Huge numbers of pilgrims/tourists visited the city at the same time that World Pride Roma was held. World Pride Roma was not a site of commercialised gay consumption; rather, the modest tents were indicative of the tensions and politics surrounding the Church and non-heteronormative sexualities. Rome is an international tourist site where some 'sacred' forms of neoliberal cultural citizenship are encouraged. The local contestation between pride supporters and the Vatican meant that as a tourist stage, Rome can be 'known' in multiple ways.

Figure 6.3 'God is Gay', World Pride Roma 2000 (source: Author's own collection).

Conclusion

This chapter examined Sydney, Auckland, Edinburgh and Rome in order to analyse international queer tourism. In many western cities the gay rights movement has insisted on justice and equality for non-heterosexual citizens. Pride parades have been a display of demands for equal rights. The politics of parades is just one function, however, and there is an increasing trend towards commodification of pride parades.

The queer spaces of the cities that I refer to in this chapter may share some common characteristics but they are also diverse and different. Cities such as Auckland and Sydney are often considered the 'jewel in the drag queen crown' of pride celebrations. Sydney, in particular, has a highly successful and commercialised month long celebration of all things queer. Mardi Gras has significant commercial sponsorship, it is promoted by the city and it is subject to state controls and regulations. Some of the parade displays may be risqué and/or politically subversive; however, pride may also be understood as a mainstream tourist event. In some cities, pride parades play important roles in the commercialisation and commodification of queer lifestyles.

Auckland's HERO Parade is situated in a 'gay' suburb – Ponsonby. Like Sydney, the parade is held along a well-defined commercialised and residential queer district. Both parades queer the streets and the empirical material suggests that these parades can be subversive of heteronormativity. Ironically, however, the promotion of difference may also contribute to a type of 'homonormativity' where queer celebrations are packaged for tourist celebrations.

Edinburgh is a city with a famous international reputation for festivals. The celebration of Pride, however, remains liminal to the city's identity and tourist reputation. Empirical material suggests that those who perform in Pride Scotland are critical of Edinburgh's conservative city politics and policies.

The celebration of queerness and Catholicism during the summer of 2000 in Rome is the final focus of the chapter. World Pride Roma was deliberately held in Rome during the Vatican's Jubilee in an attempt to open up Rome to queer tourists and to subvert the city's homophobia. Using debates that surfaced in the media, and my own experiences of World Pride Roma 2000, I stress the conflicts and continuities that arose between Catholicism and queerness.

Throughout this chapter I account for the way in which the performance of city tourist sites reflects, reinforces and resists various hegemonic power relations. By examining various city parade sites I highlight the powerful ways in which these specific spaces reflect particular and partial western discourses of embodiment and place. Parades and queer celebrations may be understood as enabling new city forms of neoliberal cultural citizenship, where queers may want to (but not all are able to) buy into it.

7 Paradoxical endings

Over three decades of pride parades mean that, collectively, a lot of people have walked a long way. Numerous outfits and costumes have been made and worn. Organising committees have co-ordinated (and cajoled) disparate groups to march under one banner. City and parade officials have had 'heated' public debates regarding 'permission' to march and under what regulations. Tourists have roared with excitement while some have been both repulsed and fascinated. Paraders have 'come out', felt proud and partied, while others may have felt excluded by the event. A central theme of this book, then, has been the contradictory and paradoxical productions of pride parades.

In this concluding chapter I draw together a number of themes that are part of a wider critical project concerned with the politics of knowledge production. Throughout this book I engage in a number of interpretive problematics concerning the relations between knowledge and sexualised subjectivities. I assume that knowledge is embodied and that bodily subjects are produced by, and positioned in, extraordinarily complex articulations of subjectivities, in particular gender and sexuality. These subjectivities are always relational; paraders are always located in relation to other paraders, tourists to other tourists, paraders to tourists, and of course, myself as a researcher to those I have researched.

In this book I place gendered and sexualised bodies at the centre of tourism research. Through detailed readings of different pride parades in different cities, I attempt to engage in the politics of sexuality, both on the street and in the academy. Tourism knowledges have, like most of the social sciences, been built upon hierarchical dualisms (Chapter 2). Notions of self/Other, tourist/host, inside/outside, centre/margin and related to this study, straight/gay, are frequently employed by social scientists to describe social relations. Such powerful positions, I have argued, do not need to be fixed. Queer tourist spaces are multi-dimensional, shifting and contingent. One way to subvert mind/body, self/Other dichotomies in tourism practices and in research is first to recognise the hierarchical relationship of self (tourist) and Other (host). The second strategy, however, is to unsettle mind/body and self/Other dichotomies. Bringing the gendered/sexed body into tourism studies epistemologically challenges the distinctions between mind and body, self and Other, tourist and host. Another implication of embodying tourism studies is that research becomes focused on groups that have been marginalised both in the academy and in tourism processes.

For the remainder of this chapter I turn to Rose's (1993b) concept of paradoxical space, in order to explore the possibilities of tourist spaces which do not replicate the static positions of self and Other, host and guest, straight and gay.

Rose (1993b: 140) argues that paradoxically we can simultaneously occupy space that is both centre and margin, inside and outside. Pride parades represent just such a paradoxical space. On the one hand, it can be positive, empowering and supportive to be involved in a pride parade. On the other hand, parades can simultaneously be a site of danger where lesbians, gay men, transgenders, bisexuals and so on can encounter a range of social risks and be subject to public abuse and social exclusion. Writing about young lesbians and gay men, Gill Valentine and Tracey Skelton (2003) note that vulnerable social groups are not just marginalised and/or oppressed, but can also marginalise and oppress each other.

The notion of paradoxical space is informed by queer theorising, in particular, de Lauretis (1987) argues that in order to break out of the masculinist (heteronormative) field of knowledge, multiple aspects of subjectivity need to be taken into account. Another way in which to challenge hegemonic and heterosexist constructions of knowledge is to go beyond the dominant forms of subjectivity. The subject of 'queer' is thus constituted:

> not by sexual difference alone, but rather across languages and cultural representations; a subject en-gendered in the experiencing of race and class, as well as sexual, relations; a subject, therefore, not unified but rather multiple, and not so much divided as contradicted.
> (de Lauretis 1987: 2)

There is subversive potential to this type of thinking. A queer critique depends on a desire for something else. Referring to themselves, Veijola and Jokinen (1994: 125) assert: 'We are not naïve. We are critical and suspicious, as scientists should be.' My suspicions are used to question and inform current tourism theorising. That tourism studies needs 'queer(y)ing' is evident in at least two ways. Early work had paid no attention to gender and sexuality, and second, tourism academics had paid no attention to embodiment: their own or the bodies of research participants. In an important analysis as to why that had been the case, Veijola and Jokinen (1994) argue that the distinction between rational and irrational had produced disembodied masculinist tourism knowledges. They ask: 'how are we going to change our research practices and tourist practices in a way that prevents us from constituting the Other *out-side* of ourselves?' (Veijola and Jokinen 1994: 148, italics in the original). Writing from one's located and embodied position is one way to unsettle positivist, rational and masculinist constructions of tourism studies researchers and I have attempted to do this throughout *Queering Tourism*.

Critical approaches to tourism are often united by a refusal to conceptualise tourism as merely involving innocent affairs of 'fun' and 'enjoyment' (Puar 2002a). Such suspicions of enjoyment are useful to illustrate how tourism facilitates, for example, labour exploitation, unequal gender relations, cultural destruction, sexual discrimination, and the perpetuation of heteronormativity. These critical approaches are not critical enough insofar as they tacitly assume that tourism enjoyment is just enjoyment and easily enjoyed.

As a noun and a verb 'queer' is used to create uncertainty and disjuncture between gender, sexuality and desire (Jagose 1996). For too long heteronormative understandings of embodied subjectivities have been held up as the norm. Dominant forms of femininity and masculinity have assumed 'natural' status within the heterosexual nexus. I employ theoretical notions of abjection, camp and performativity to make sense of the paradoxical spaces of gay pride parades as tourist events.

Starting with the body, then, Chapter 3 turns towards a camp epistemology (Binnie 1997) in order to examine two types of parade groups: a women's marching/drumming group and male marching groups. These similar but different marching performances are contingent upon, and constituted by the tourist spaces they occupy. The specificities of parade sites – Sydney, Auckland, and Edinburgh – are crucial to the production of camp performances. Marching girls in Edinburgh use camp sensibilities in order to enjoy Pride Scotland and to dispel indifferent or hostile tourist reactions. In a city that has a reputation for being conservative, Edinburgh's marching girls find themselves in an 'accidental' tourist event. Spectators are unsure of their roles and the encounter between paraders and tourists produces ambivalent and paradoxical space. At the other 'end of the world', marching boys in Auckland and Sydney rely on global discourses of queer tourism which in turn 'prepare' tourist stages. Marching boys' camped up performances – where gendered binaries of masculine and feminine are parodied and disrupted (Butler 1990, 1993) – make tourists roar with excitement. Sydney and Auckland become neoliberal cities of queer consumption. The paradox of marching groups rests on the difference that space makes to camp performances.

Drawing on the notion of abjection (Kristeva 1982) I put forward the idea that queer bodies may personify the border between self and Other (Chapter 4). These borders come loose, however, despite efforts to secure them. 'Loose borders' may both excite and repulse various tourists at pride parades. Furthermore, particular bodies and performances that threaten the borders of corporeal acceptability also may threaten the heteronormative space of streets. The more flamboyant and camp the performance, the more popular the tourist event. This may be due to the ability for straight tourists to visually distance themselves from the bodies on display. When the queer body can't be distinguished from the straight body, however, border anxiety is heightened. Some bodies involved in risqué displays of bondage and sadomasochism (s/m), bodies living with HIV/AIDS, and bodies which contest the gendered corporeal borders, may be constructed as 'freaks', 'ugly', 'dirty' (I. M. Young 1990) and are Othered by particular watching tourists. Those bodies which perform the 'norm', such as the high school group 'Rainbow Youth' may work to dissolve the conceptual border between gay and straight, self and Other.

Turning to a deconstructive analysis of western hierarchical dualisms helps uncovers some of the complexities and power relations involved in tourism processes. Notions of what is deemed appropriate in public and 'private' spaces structures the analysis of Chapter 5: 'Sex in the suburbs or the CBD?' The decision to move Auckland's HERO Parade from down town Auckland to the 'gay' suburb Ponsonby was debated in New Zealand media for several months. This debate highlights the spatial paradoxes of gay pride parades. HERO in Queen Street came to represent resistance to heteronormative city spaces. A march down the main street of Auckland, New Zealand's biggest city, stood for protest and calls for equal rights. HERO in Ponsonby was represented as a 'party in our own neighbourhood', and hence deemed politically ineffectual. Neither of these positions is fixed; rather, politics and celebrations pervade both spaces, but in different, powerful and sometimes hegemonic ways. Resistance to HERO being held in any part of the city still exists, as well as acceptance of HERO as a carnival type event in Ponsonby.

The pleasures and 'successes' of queer consumption, that is, the incorporation of queer celebrations into city promotions and mainstream media, may be understood as a kind of 'homonormativity'. Chapter 6 explores the contradictions surrounding the 'acceptance' of gay pride parades and their adoption of neoliberal commercial policies and practices. In some cities, such as Sydney, Auckland, San Francisco, transgressive political displays are now corporatised, regulated and controlled. Debates on urban governance and sexual citizenship inform the construction of Sydney's – and to some extent Auckland's – queer tourist space.

As this book has shown, in some places, at particular times, the idea of homosexuality as culturally subversive is a popularly held notion. Queer bodies may be understood as threats to social stability and public order. Further to this, then, has been the perceived need of external control and regulation. In recent years, rather than critique social institutions and practices that have historically excluded queers, the dominant discourse of the last decade has been that of seeking access to mainstream culture through demanding equal rights (Richardson 2005). Paraders may want to 'buy into' – integrate and assimilate – a new form of neoliberal cultural citizenship.

'Normal' and 'ordinary' and 'good' gay citizens, however, do not dominate the political landscapes of cities such as Edinburgh and Rome.

Alongside Sydney and Auckland, Chapter 6 also considers the paradoxical tourist space of Pride Scotland in Edinburgh. Here participants critique the conservative (homophobic and racist) construction of Edinburgh. Internationally famous for its festivals, Edinburgh's Pride Scotland, however, confuses tourists as it is outside 'mainstream' publicity networks. The political potential in this confusion is that it may subvert new forms of sexualised neoliberal citizenship.

A similar paradox existed in Rome when World Pride Roma 2000 was held alongside *Giubileo* – the celebrations of 2000 years of Christianity. Organisers of World Pride Roma had hoped that the event would constitute Rome as a cosmopolitan queered city. Public debates were played out in the media between the Vatican, city officials and parade organisers. During World Pride the 'sacred city' of Rome did not become defined by gay tourism commercialisation. Rather, it became an exhibition space for a commercially and spiritually motivated Church. Rome is an international tourist site where some 'sacred' forms of neoliberal cultural citizenship are encouraged.

In concluding I want to make a series of points in relation to the ideas discussed in this book. My first point is that it is not enough to examine only the global scales of tourism processes. The politics of sexualised bodies pervade multiple spaces. The chapters in this book are loosely grouped around different scales – body, street, suburb and city – in order to illustrate that micro- and macro-level tourist spaces are always sexed and sexualised.

The second point I wish to make is that knowledge is always political. My hope is that this research challenges western constructions of disembodied heteronormative knowledge. Rather than just adding the body to tourism studies, the use of queer theories on the body challenges existing conceptions of tourism events. Using these ideas to understand the power relationships between tourists and the participants of the gay pride parades has led to a redefinition of tourism processes. Attention to the (gendered, sexualised, raced/coloured, (dis)abled and so forth) embodied processes of tourist events and practices has much potential and there is room for much more research in this area. It then becomes possible to theorise 'embodiment, radical Otherness, multiplicity of differences, sex and sexuality in tourism' (Veijola and Jokinen 1994: 129). Binnie and Valentine (1999: 183) write: 'Geographers could occupy a central place in the rearticulation of queer theory to include a much needed social or material dimension'. This book has attempted to mesh queer theories and social 'realities'.

The third and final point I want to make is that by understanding tourism geographies as paradoxical performances offers an opportunity to undo or destabilise sexualised categories that are problematic. Butler (in Mahtani 2001: 302) states that:

we are never fully determined by the categories that construct us, but yet at the same time, one can dissent from the norms of society by occupying the very categories by which one is constituted and turning them in another direction or giving them a future they weren't supposed to have.

It is possible to think of the Other as not outside of the Same. Positions can be occupied simultaneously: inside-ness and outside-ness produce unpredictable paradoxes. In examining gay pride performances through the scales of body, street, suburb and city, I have illustrated that what all these tourist spaces share is their occupation of both centre and margin. Although some bodies are commonly represented as more abject or Other, in fact all bodies (particularly heteronormative tourists' bodies) are unstable, contingent and paradoxical.

The queered tourism spaces of gay pride parades challenge some of the dominant constructions of knowledge in tourism studies. I want to end by calling for disciplinary futures that are uncertain, unstable, shifting and always contested. Queerness contests heteronormativity, a politics of pride makes precarious a politics of shame, and paradoxical bodies unsettle firm, bounded and exclusionary knowledges.

References

Adam, Barry (1995) *The Rise of the Gay and Lesbian Movement*, Toronto: Prentice Hall.
Aldrich, Robert (1993) *The Seduction of the Mediterranean: Writing, Art and Homosexual Fantasy*, London: Routledge.
Antipode (2002) Special Issue 'Queer Patriarchies, Queer Racisms, International', 34, 5: 835–987.
Annals of Tourism Research: A Social Science Journal (1995) 22, 2: 247–489.
Armour, Lawrence (2003) 'Bridle accessories', *Fortune*, October 13, 148, 8: 60.
Australian Broadcasting Corporation Video (1995) *Sydney Mardi Gras Parade* (programme transcript), ABC Video: Sydney.
Baudrillard, Jean (1988) *America*, London: Verso.
Becker, Toby (2000) 'Rome's Mayor defends plan to host gay pride gathering', *National Catholic Reporter*, 11 February, 36, 15: 9.
Bell, David (1995) '[scew]ing geography (censor's version)', *Environment and Planning D: Society and Space* 13: 127–131.
Bell, David and Binnie, Jon (2000) *The Sexual Citizen: Queer Politics and Beyond*, Cambridge: Polity Press.
Bell, David; Binnie, Jon; Cream, Julia and Valentine, Gill (1994) 'All hyped-up and no place to go', *Gender, Place and Culture: A Journal of Feminist Geography*, 1: 31–48.
Bell, David and Jayne, Mark (eds) (2004) *City of Quarters: Urban Villages in the Contemporary City*, Aldershot: Ashgate.
Bell, David and Valentine, Gill (eds) (1995) *Mapping Desires: Geographies of Sexualities*, London: Routledge.
Bell, David and Valentine, Gill (eds) (1997) *Consuming Geographies: You Are Where You Eat*, London: Routledge.
Bennie, J. (1995) 'Warriors court marching boys', *Man to Man: New Zealand's National Gay Community Newsletter*, 13 April: 1.
Berg, Larry (1994) 'Masculinity, place, and a binary discourse of "Theory" and empirical investigation in the human geography of Aotearoa/New Zealand', *Gender, Place and Culture: A Journal of Feminist Geography*, 1: 245–260.
Binnie, Jon (1995) 'Trading places: Consumption, sexuality and the production of queer space', in David Bell and Gill Valentine (eds) *Mapping Desires: Geographies of Sexualities*, London: Routledge, pp. 182–199.
Binnie, Jon (1997) 'Coming out of Geography: Towards a queer epistemology?', *Environment and Planning D: Society and Space*, 15: 223–237.

Binnie, Jon (2004) 'Quartering sexualities: Gay villages and sexual citizenship', in David Bell and Mark Jayne (eds) (2004) *City of Quarters: Urban Villages in the Contemporary City*, Aldershot: Ashgate, pp. 163–172.

Binnie, Jon and Valentine, Gill (1999) 'Geographies of sexualities – a review of progress', *Progress in Human Geography*, 23: 175–187.

Blum, Virginia (2002) 'Introduction: The liberation of intimacy: Consumer-object relations and (hetero)patriarchy, *Antipode*, 34, 5: 845–863.

Bondi, Liz (1998) 'Sexing the city', in Jane Jacobs and Ruth Fincher (eds) *Cities of Difference*, New York and London: The Guilford Press, pp. 177–200.

Bondi, Liz; Avis, Hannah; Bingley, Amanda; Davidson, Joyce; Duffy, Rosaleen; Einagel, Victoria; Green, Anja-Maaike; Johnston, Lynda; Lilley, Sue; Listerborn, Corinna; Marshy, Mona; McEwan, Shonagh; O'Connor, Niamh; Rose, Gillian; Vivat, Bella; Wood, Nichola (2002) *Subjectivities, Knowledges and Feminist Geographies: The Subjects and Ethics of Social Research*, Lanham, Maryland: Rowman and Littlefield.

Bonnemaison, S. (1990) 'City policies and cyclical events', *Celebrations: Urban Spaces Transformed, Design Quarterly*, 147, Cambridge, Massachusetts, Institute of Technology for the Walker Art Center: 24–32.

Bordo, Susan (1986) 'The Cartesian masculinisation of thought', *Signs*, 11: 239–256.

Bordo, Susan (1992) 'Review essay: Postmodern subjects, postmodern bodies', *Feminist Studies*, 18: 159–175.

Brickell, Chris (2000) 'Heroes and invaders: Gay and lesbian pride parades and the public/private distinction in New Zealand media accounts', *Gender, Place and Culture: A Journal in Feminist Geography*, 7, 2: 168–178.

Brighton Ourstory Project (1992) *Daring Hearts*, Brighton: QueenSpark Books.

Brimble, D. (1995) 'Stick with Queen Street', *express: new zealand's newspaper of gay expression*, 28 September: 2.

Britton, Steven (1991) 'Tourism, capital and place: Towards a critical geography of tourism', *Environment and Planning D: Society and Space*, 9, 4: 451–478.

Brown, Michael (1995) 'Ironies of Distance: An Ongoing Critique of the Geographies of AIDS', *Environment and Planning D: Society and Space*, 13: 159–183.

Brown, Michael (2000) *Closet Space: Geographies of Metaphors from the Body to the Globe*, London: Routledge.

Bruto, L. G. (1997) *Cicciobello del Potere*, Milan: Kaos.

Butler, Judith (1990) *Gender Trouble: Feminism and the Subversion of Identity*, New York and London: Routledge.

Butler, Judith (1993) *Bodies that Matter: On the Discursive Limits of Sex*, New York and London: Routledge.

Cady, Joseph (1992) '"Masculine love", Renaissance writing and the "new invention" of homosexuality', *Journal of Homosexuality*, 23, 1–2: 9–40.

Caldwell, F. (1995) 'Petitions both ways', *express: new zealand's newspaper of gay expression*, 12 October 1995: 3.

Callard, Felicity (1998) 'The body in theory,' *Environment and Planning D: Society and Space*, 16: 387–400.

Cameron, E. (1995) 'Petition seeks to retain Queen St. parade route', *express: new zealand's newspaper of gay expression*, 14 September 1995: 3.

Carbery, G. (1995) *A History of the Sydney Gay and Lesbian Mardi Gras*, Parkville, Victoria: Australian Lesbian and Gay Archives Inc.

Castaneda, Q. (1991) 'An "Archaeology" of Chichen Itza: Discourse, Power and Resistance in a Maya Tourist Site'. Unpublished doctoral dissertation at the State University of New York at Albany.

Chouinard, Vera and Grant, Ali (1996) 'On not even being anywhere near the project: Putting ourselves in the picture', in Nancy Duncan (ed.) *BodySpace: Destabilising Geographies of Gender and Sexuality*, London and New York: Routledge, pp. 170–196.

Christie, Nigel (1995) 'Parade politics', *express: new zealand's newspaper of gay expression*, 26 October: 2.

Cieri, Marie (2003) 'Between being and looking: Queer tourism promotion and lesbian social space in Greater Philadelphia', *ACME: An International E-Journal for Critical Geographies*, 2, 2: 147–166.

Cleto, Fabio (ed.) (1999) *Camp: Queer Aesthetics and the Performing Subject: A Reader*, Edinburgh: Edinburgh University Press.

Clift, Stephen and Carter, Simon (eds) (2000) *Tourism and Sex: Culture, Commerce and Coercion*, London and New York: Pinter.

Cloke, Paul and Perkins, Harvey (1998) '"Cracking the canyon of the awesome foursome": Representations of adventure tourism in New Zealand', *Environment and Planning D: Society and Space*, 16: 185–218.

Cohen, A. (1980) 'Drama and politics in the development of a London carnival', *Man*, (n.s) 15: 65–87.

Cohen, A. (1982) 'A polyethnic London carnival as a contested cultural performance', *Ethnic and Racial Studies*, 5, 1: 23–41.

Cohen, E. (1988) 'Tourism and AIDS in Thailand', *Annals of Tourism Research: A Social Science Journal*, 15: 467–486.

Cohen, E. (1995) 'Contemporary tourism: Trends and challenges. Sustainable authenticity or contrived post-modernity?', in R. W. Butler and D. G. Pearce (eds) *Change in Tourism: People, Places and Processes*, London and New York: Routledge, pp. 12–29.

Cohen, Roger (2000) 'Night moves of all kinds: Berlin', *New York Times*, 17 September, travel section: 11.

Compact Oxford English Dictionary (second edition) (1991) Oxford: Clarendon Press.

Craik, Jennifer (1997) 'The culture of tourism', in Chris Rojek and John Urry (eds) *Touring Cultures: Transformations of Travel and Theory*, London and New York: Routledge, pp. 113–136.

Cream, Julia (1992) 'Sexing shapes', *Sexuality and Space Network, Lesbian and Gay Geographers? Proceedings*, London.

Cream, Julia (1995) 'Re-solving riddles: The sexed body', in David Bell and Gill Valentine (eds) *Mapping Desires: Geographies of Sexualities*, London: Routledge, pp. 31–40.

Cresswell, Tim (1996) *In Place/Out of Place: Geography, Ideology and Transgression*, Minneapolis, University of Minnesota.

Dahir, Mubarak (2001) 'Pride at work', *The Advocate*, June 19: 38.

De Lauretis, Teresa (1987) *Technologies of Gender: Essays on Theory, Film and Fiction*, London: Macmillan.

De Lauretis, Teresa (1991) 'Queer theory: lesbian and gay sexualities: an introduction', *differences*, 3: 1–10.

Del Casino, Vincent and Hanna, Stephen (2000) 'Representations and identities in tourism map spaces', *Progress in Human Geography*, 24, 1: 23–46.

Derrida, Jacques (1981) *Dissemination* (trans. B. Johnson). Chicago: University of Chicago Press.
Doane, M. (1990) 'Film and the masquerade: Theorizing the female spectator', in P. Erens (ed.) *Issues in Feminist Film Criticism*, Bloomington: Indiana University Press, pp. 41–57.
Douglas, Mary (1980) *Purity and Danger: An Analysis of the Concepts of Pollution and Taboo*, London: Routledge and Kegan Paul.
Duncan, Nancy (1996) 'Renegotiating gender and sexuality in public and private spaces', in Nancy Duncan (ed.) *BodySpace: Destabilizing Geographies of Gender and Sexuality*, London and New York: Routledge, pp. 127–145.
Dyer, Richard (1993) 'Straight acting', in Richard Dyer (ed.) *The Matter of Images: Essays on Representations*, London: Routledge, pp. 133–136.
Dyer, Richard (2002) *The Culture of Queers*, London and New York: Routledge.
Eagleton, T. (1981) *Walter Benjamin: Towards a Revolutionary Criticism*, London: Verso.
Eco, U. (1986) *Travels in Hyper-reality*, London: Picador.
Economist (2002) 'The geography of cool', 15 April: 91.
Edensor, Tim (2001) 'Performing tourism, staging tourism: (Re)producing tourist space and practice', *Tourist Studies*, 1: 59–81.
Edensor, Tim and Kothari, Uma (1994) 'The masculinisation of Stirling's heritage', in Vivian Kinnaird and Derek Hall (eds) *Tourism: A Gendered Analysis*, Chichester: John Wiley and Sons, pp. 164–187.
Ely, L. (1996) 'Sex billboard sparks stir', *New Zealand Herald*, 9 February: 2.
Enloe, Cynthia (1989) *Bananas, Beaches and Bases: Making Feminist Sense of International Politics*, London: Pandora.
Environment and Planning D: Society and Space (2000) Special issue on performativity, 18.
express: new zealand's newspaper of gay expression (1995) 26 October: 17.
express: new zealand's newspaper of gay expression (1995) 7 December: 6.
express: new zealand's newspaper of gay expression (1995) 9 November: 10.
express: new zealand's newspaper of gay expression (1996) 1 February: 13.
express: new zealand's newspaper of gay expression (1996) 15 February: 7.
express: new zealand's newspaper of gay expression (1998) 21 February.
Eyewitness News (1994) TV2, Newstel Log, Wednesday 2 March.
Featherstone, M. (1991) *Consumer Culture and Postmodernism*, London: Sage.
Feifer, M. (1985) *Going Places*, London: Macmillan.
Fisher, A., Hatch, J. and Paix, P. (1986) 'Road accidents and the Grand Prix', in J. P. A. Burns and T. J. Mules (eds) *The Adelaide Grand Prix: The Impact of a Special Event*, Centre for South Australian Economic Studies: Adelaide, pp. 151–168.
Foucault, Michel (1970) *The Order of Things: An Archaeology of the Human Sciences*, London: Tavistock.
Foucault, Michel (1976) *Discipline and Punish: The Birth of the Prison*, New York: Vintage.
Foucault, Michel (1981) *The History of Sexuality: Volume One: An Introduction*, Trans. R. Hurley, London: Penguin Books Ltd.
Fox, T. C. (1995) *Sexuality and Catholicism*, New York: George Braziller.
Game, Ann (1991) *Undoing the Social: Towards a Deconstructive Sociology*, Buckingham: Open University Press.
GLQ: A Journal of Lesbian and Gay Studies (2002) 'Queer tourism: Geographies of globalization', 8: 1–2.

Gearing, Nigel (1994) 'Hero v homebody: A gay debate', *New Zealand Listener*, November 26 (147): 26–27.

Gearing, Nigel (1997) *Emerging Tribes: Gay Culture in New Zealand in the 1990s*, Auckland: Penguin.

Giorgi, Gabriel (2002) 'Madrid en tránsito: Travelers, visibility, and gay identity', *GLQ: A Journal of Lesbian and Gay Studies*, 8, 1–2: 57–79.

Greenwood, D. (1978) 'Culture by the pound: An anthropological perspective on tourism as cultural commodification', in Valene Smith (ed.) *Hosts and Guests. The Anthropology of Tourism*, Oxford: Blackwell.

Gregory, Derek (1994) *Geographical Imaginations*, Cambridge: Blackwell.

Gregson, Nicky and Rose, Gillian (2000) 'Taking Butler elsewhere: Performativities, spatialities and subjectivities', *Environment and Planning D: Society and Space*, 18 (4): 433–452.

Grosz, Elizabeth (1989) *Sexual Subversions: Three French Feminists*, Sydney: Allen and Unwin.

Grosz, Elizabeth (1992) 'Bodies-Cities', in Beatrice Colomina (ed.) *Sexuality and Space*, New York: Princeton University Press, pp. 241–253.

Grosz, Elizabeth (1993) 'Bodies and knowledges: Feminism and the crisis of reason', in L. Alcoff and E. Potter (eds) *Feminist Epistemologies*, New York: Routledge, pp. 187–215.

Grosz, Elizabeth (1994) *Volatile Bodies: Towards a Corporeal Feminism*, St Leonards: Allen and Unwin.

Grosz, Elizabeth (1995) *Space, Time and Perversion: The Politics of Bodies*, Sydney: Allen and Unwin.

Hall, C. Michael (1992) *Hallmark Tourist Events*, Belhaven Press: London.

Haraway, Donna (1991) *Simians, Cyborgs and Women: The Reinvention of Nature*, London: Free Association Books.

Harvey, David (1989) *The Condition of Postmodernity*, Oxford: Blackwell.

HERO Magazine (1996) Auckland: The HERO Project Limited.

Hill, Kim (1994) 'Kim Hill, 9 to Noon Show', *Radio New Zealand*, Wednesday, 13 April.

Hill, K. (1998) 'Record crowd forms a sea of celebration'. *The Sydney Morning Herald*, http://www.smh.com.au/daily/content/980302/national1.html (accessed 2 May 1998).

Hodge, Stephen (1995) 'No fags out there: Gay men, identity and suburbia', *Journal of Interdisciplinary Gender Studies*, 1: 41–8.

Holcomb, B. and Luongo, M. (1996) 'Gay tourism in the United States', *Annals of Tourism Research: A Social Science Journal*, 23, 3: 711–712.

Holmes, Paul (1994), 'Paul Holmes', *Radio New Zealand 1ZB Newstel Log*, 11 March, Auckland.

Hubbard, Phil; Kitchin, Rob; Bartley, Brendan and Fuller, Duncan (2002) *Thinking Geographically: Space, Theory and Contemporary Human Geography*, London: Continuum.

Hughes, Howard (1997) 'Holidays and homosexual identity', *Tourism Management* 18, 1: 3–7.

Hughes, Howard (1998) 'Sexuality, tourism and space', in D. Tyler, M. Robertson and Y. Guerrier (eds) *Managing Tourism in Cities: Policy, Process and Practice*, Chichester: Wiley, pp. 163–178.

Jackson, Peter (1988) 'Street life: The politics of carnival', *Environment and Planning D: Society and Space*, 6, 2: 213–227.

Jackson, Peter (1991) 'The cultural politics of masculinity: Towards a social geography', *Transactions of the Institute of British Geographers*, 16: 199–213.
Jackson, Peter (1992) 'The politics of the streets: A geography of Caribana', *Political Geography*, 11: 1–22.
Jackson, Peter (1994) 'Black male: Advertising and the cultural politics of masculinity', *Gender, Place and Culture: A Journal of Feminist Geography*, 1, 1: 49–59.
Jagose, Annamarie (1996) *Queer theory*, Dunedin: University of Otago Press.
James, Bev and Saville-Smith, Kay (1994) *Gender, Culture and Power: Changing New Zealand's Gendered Culture*, Auckland: Oxford University Press.
Johnson, Louise C. (1989) 'Embodying geography: Some implications for considering the sexed body in space', *New Zealand Geographical Society Proceedings of the 15th New Zealand Geography Conference*, Dunedin: 134–138.
Johnston, Lynda (1995) 'The politics of the pump: Hard core gyms and women body builders', *New Zealand Geographer*, 51, 1: 16–18.
Johnston, Lynda (1996) 'Pumped up politics: Female body builders refiguring the body', *Gender, Place and Culture: A Journal of Feminist Geography*, 3, 3: 327–340.
Johnston, Lynda (1997) 'Queen(s') Street or Ponsonby poofters? The embodied HERO Parade site', *New Zealand Geographer*, 53: 29–33.
Johnston, Lynda (1998) 'Reading sexed bodies in sexed spaces', in Heidi Nast and Steve Pile (eds) *Places Through the Body*, London: Routledge, pp. 244–262.
Johnston, Lynda (2001) '(Other) Bodies and tourism studies', *Annals of Tourism Research: A Social Science Journal*, 28: 180–201.
Johnston, Lynda (2002) 'Borderline bodies at gay pride parades', in Bondi, Liz; Avis, Hannah; Bingley, Amanda; Davidson, Joyce; Duffy, Rosaleen; Einagel, Victoria; Green, Anja-Maaike; Johnston, Lynda; Lilley, Sue; Listerborn, Corinna; Marshy, Mona; McEwan, Shonagh; O'Connor, Niamh; Rose, Gillian; Vivat, Bella; Wood, Nichola, *Subjectivities, Knowledges and Feminist Geographies. The Subjects and Ethics of Social Research*, Lanham, Maryland: Rowman and Littlefield, pp. 75–89.
Johnston, Lynda (2005) 'Man: Woman', in Paul Cloke and Ron Johnston (eds) *Spaces of Geographical Thought*, London: Sage, pp. 119–141.
Johnston, Lynda and Valentine, Gill (1995) 'Where ever I lay my girlfriend that's my home: Performance and surveillance of lesbian identity in home environments', in David Bell and Gill Valentine (eds) *Mapping Desires: Geographies of Sexualities*, London: Routledge, pp. 99–113.
Jokinen, Eeva and Veijola, Soile (1997) 'The disoriented tourist: The figuration of the tourist in contemporary cultural critique', in Chris Rojek and John Urry (eds) *Touring Cultures: Transformations of Travel and Theory*, London and New York: Routledge, pp. 23–51.
Kantsa, Venetia (2002) '"Certain places have different energy": Spatial transformations in Eresos, Lesvos', *GLQ: A Journal of Lesbian and Gay Studies*, 8, 1–2: 35–55.
Kates, Steven M. and Belk, Russell W. (2001) 'The meanings of lesbian and gay pride day', *Journal of Contemporary Ethnography*, 30, 4: 392–429.
Keith, H. (1994) 'Along the strip', *The Strip*, April: 19–20.
Kinnaird, Vivian and Hall, Derek (eds) (1994) *Tourism: A Gender Analysis*, Chichester: John Wiley and Sons Ltd.
Kippendorf, J. (1987) *The Holiday Makers: Understanding the Impact of Leisure and Travel*, Oxford: Heinemann Professional Publishing.
Kirby, Andrew (1995) 'Straight talk on the pomo question', *Gender, Place and Culture: A Journal of Feminist Geography*, 2: 89–96.

Kirby, Vicky (1997) *Telling Flesh: The Substance of the Corporeal*, New York and London: Routledge.
Knopp, Lawrence (1995) 'If you're going to get all hyped up you'd better go somewhere!', *Gender, Place and Culture: A Journal of Feminist Geography*, 2: 85–88.
Knopp, Lawrence (1998) 'Sexuality and urban space: Gay male identity politics in the United States, the United Kingdom, and Australia', in Jane Jacobs and Ruth Fincher (eds) *Cities of Difference*, New York and London: The Guilford Press, pp. 149–176.
Kristeva, Julia (1982) *Powers of Horror: An Essay on Abjection*. Trans. Leon S. Roudiez, New York: Columbia University Press.
La Rocca, O. (2000) 'La manifestazione gay un affronto al giubileo', *La Repubblica*, 10 June: 2.
Legat, Nicola (1994) 'Hay fever', *Metro*, February: 86–96.
Levine, M. (1987) 'Downtown redevelopment as an urban growth strategy: A critical appraisal of the Baltimore renaissance', *Journal of Urban Affairs*, 9, 2: 103–123.
Lewis, Claire and Pile, Steve (1996) 'Woman, body, space: Rio Carnival and the politics of performance', *Gender, Place, and Culture: A Journal of Feminist Geography*, 3: 23–41.
Ley, D. and Olds, K. (1988) 'Landscape as spectacles: World's fairs and the culture of heroic consumption', *Environment and Planning D: Society and Space*, 6, 2: 191–212.
Lloyd, Genevieve (1993) *The Man of Reason: 'Male' and 'Female' in Western Philosophy*, London: Routledge.
Longhurst, Robyn (1995) 'Geography and the body', *Gender, Place and Culture: A Journal of Feminist Geography*, 2: 97–105.
Longhurst, Robyn (1997) '(Dis)embodied geographies', *Progress in Human Geography*, 21, 4: 486–501.
Longhurst, Robyn (2001a) *Bodies: Exploring Fluid Boundaries*, London and New York: Routledge.
Longhurst, Robyn (2001b) 'Classics in human geography revisited', *Progress in Human Geography*, 25, 2: 252–255.
Luongo, Michael (2002) 'Rome's world pride', *GLQ: A Journal of Lesbian and Gay Studies*, 8, 1–2: 167–181.
MacCannell, Dean (1989) *The Tourist: A New Theory of the Leisured Class*, New York: Schocken Books.
MacCannell, Dean (1992) *Empty Meeting Grounds: A New Theory of the Leisured Class*, London: Routledge.
McClintock, Anne (1995) *Imperial Leather: Race, Gender and Sexuality in the Colonial Conquest*, New York: Routledge.
McDowell, Linda (1991) 'The baby and the bath water: Diversity, deconstruction and feminist theory', *Geoforum*, 22: 123–133.
McNeill, Donald (2003) 'Rome, global city? Church, state and the Jubilee 2000', *Political Geography*, 22: 535–556.
Mahtani, Minelle (2001) 'Racial remappings: The potential of paradoxical space', *Gender, Place and Culture: A Journal of Feminist Geography*, 8, 3: 299–305.
Manning, F. E. (1989) 'Carnival in the city: The Carribeanization of urban landscapes', *Urban Resources*, 5, 3: 3–43.
Markwell, Kevin (2002) 'Mardi Gras tourism and the construction of Sydney as an

international gay and lesbian city', *GLQ: A Journal of Lesbian and Gay Studies*, 8, 1–2: 81–99.
Marshment, M. (1997) 'Gender takes a holiday', in Margaret T. Sinclair (ed.) *Gender, Work and Tourism*, New York: Routledge, pp. 16–34.
Massey, Doreen (1993) 'Power-geometry and a progressive sense of place', in J. Bird *et al.* (eds) *Mapping the Futures: Local Cultures, Global Change*, London: Routledge, pp. 59–69.
Massey, Doreen (1994) *Space, Place and Gender*, Cambridge: Polity Press.
Miles, M. and Huberman, A. M. (1994) *Qualitiative Data Analysis: An Expanded Source Book*, Thousand Oaks: Sage.
Moore, L. (1996) 'New route wins hearts of Hero Parade', *New Zealand Herald*, 19 February: 3.
Morin, Karen; Longhurst, Robyn; and Johnston, Lynda (2001) '(Troubling) Spaces of mountains and men: New Zealand's Mount Cook and Hermitage Lodge', *Social and Cultural Geography*, 2, 2: 117–139.
Morris, Meghan (1988a) 'Things to do with shopping centres', in S. Sheridan (ed.) *Grafts: Feminist Cultural Criticism*, London: Verso, pp. 193–225.
Morris, Meghan (1988b) 'At Henry Parks Motel', *Cultural Studies*, 2: 1–47.
Mulvey, Laura (1975) 'Visual pleasures and narrative cinema', *Screen*, 16, 3: 6–19.
Munt, Ian (1994) 'The 'Other' postmodern tourism: Culture, travel and the new middle classes', *Theory, Culture and Society*, 11: 101–123.
Munt, Sally (1995) 'The lesbian flâneur', in David Bell and Gill Valentine (eds) *Mapping Desire: Geographies of Sexualities*, London: Routledge, pp. 114–125.
Munt, Sally (1998) *Heroic Desire: Lesbian Identity and Cultural Space*, London: Cassell.
Myslik, Wayne (1996) 'Renegotiating the social/sexual identities of places', in Nancy Duncan (ed.) *BodySpace: Destabilizing Geographies of Gender and Sexualities*, London and New York: Routledge, pp. 156–169.
Namaste, Kim (1992) 'If you're in clothes, you're in drag', *Fuse*, Fall: 7–9.
Namaste, Kim (1996) 'Gender bashing: Sexuality, gender, and the regulation of public space', *Environment and Planning D, Society and Space*, 14: 221–240.
Nast, Heidi (2002) 'Queer patriarchies, queer racisms, international. Prologue: Crosscurrents', *Antipode*, 34, 5: 835–844.
Nast, Heidi and Pile, Steve (eds) (1998) *Places Through the Body*, London and New York: Routledge.
Nelson, Lise (1999) 'Bodies (and spaces) do matter: The limits of performativity', *Gender, Place and Culture: A Journal of Feminist Geography*, 6: 331–353.
New Zealand Herald (1994) 21 February: 10.
New Zealand Herald (1994) 2 March: 2.
New Zealand Herald (1995) 'Harold Angel's week's end – It's getting worse folks', 19 August: 7.
New Zealand Herald (1997) 'Marcho men', 24 February: 20.
Nietzsche, Frederick (1967) *The Will to Power* (trans. W. Kaugmann and R. J. Hollingdale), New York: Random House.
Nietzsche, Frederick (1969) *On the Geneology of Morals* (trans. W. Kaugmann and R. J. Hollingdale), New York: Vintage.
North Shore Advertiser (1995) 10 August: 7.
Olds, K. (1988) 'Planning for the Housing Impacts of a Hallmark Event: A Case Study of Expo 1986'. Unpublished MA Thesis. School of Community and Regional Planning, University of Columbia: Vancouver.

Oliver, Kelly (1993) *Reading Kristeva: Unravelling the Double Bind*, Bloomington: Indiana University Press.

Peace, Robin; Longhurst, Robyn; and Johnston, Lynda (1997) 'Producing feminist geography "down under"', *Gender, Place and Culture: A Journal of Feminist Geography*, 4, 1: 115–119.

Phelan, Shane (2001) *Sexual Strangers. Gays, Lesbians and Dilemmas of Citizenship*, Philadelphia: Temple University Press.

Phillips, Jock (1996) *A Man's Country? The Image of the Pakeha Male, A History*, Auckland: Penguin Books.

Piggford, George (1999) 'Who's that girl? Annie Lennox, Woolf's Orlando, and female camp androgyny', in Fabio Cleto (ed.) *Camp: Queer Aesthetics and the Performing Subject: A Reader*, Edinburgh: Edinburgh University Press, pp. 283–299.

Podmore, Julie (2001) 'Lesbians in the crowd: Gender, sexuality and visibility along Montreal's Boul. St-Laurent', *Gender, Place and Culture: A Journal of Feminist Geography*, 8, 4: 333–355.

Pogglioli, Sylvia (2000) 'Profile: Italy and Vatican take opposing stands on Gay Pride festivities scheduled for July', *National Public Radio Transcript*, Washington D.C.: 30 May.

Pritchard, A.; Morgan, N.; Sedgely, D. and Jenkins, A. (1998) 'Reaching out to the gay tourist: Opportunities and threats in an emerging market segment', *Tourism Management*, 19, 3: 273–282.

Probyn, Elspeth (1993) *Sexing the Self: Gendered Positions in Cultural Studies*, London: Routledge.

Probyn, Elspeth (1995) 'Lesbians in space: Gender, sex and the structure of missing', *Gender, Place and Culture: A Journal of Feminist Geography*, 2: 77–84

Puar, Jasbir Kaur (ed.) (2002a) 'Introduction', *GLQ: A Journal of Lesbian and Gay Studies*, 8, 1–2: 1–6.

Puar, Jasbir Kaur (2002b) 'Circuits of queer mobility: Tourism, travel and globalisation', *GLQ: A Journal of Lesbian and Gay Studies*, 8, 1–2: 101–138.

Puar, Jasbir Kaur (2002c) 'A transnational feminist critique of queer tourism', *Antipode: A Radical Journal of Geography*, 34, 5: 935–946.

Richardson, Diane (2000) *Rethinking Sexuality*, London: Sage.

Richardson, Diane (2005) 'Desiring Sameness? The rise of neo-liberal politics of Normalisation', *Antipode: A Radical Journal of Geography*, 37, 3: 515–535.

Roche, M. (1990) *Mega-Events and Mico-Modernization: On the Sociology of New Urban Tourism*, Sheffield: Policy Studies Centre, University of Sheffield.

Roche, M. (1991) *Mega-Events and Urban Policy: A Study of Sheffield's World Student Games*, Sheffield: Policy Studies Centre, University of Sheffield.

Rogers, Warwick (1998) 'Columnist rejects homophobic label', *Waikato Times*, 26 June: 6.

Rojek, Chris (1993) *Ways of Escape: Modern Transformation of Leisure and Travel*, Cambridge: Polity Press.

Rojek, Chris (1997) 'Indexing, dragging, and the social construction of tourist sights', in Chris Rojek and John Urry (eds) *Touring Cultures: Transformations of Travel and Theory*, London and New York: Routledge, pp. 52–74.

Rose, Gillian (1991) 'On being ambivalent: Women and feminisms in geography', in Chris Philo (ed.) *New Words, New Worlds: Reconceptualising Social and Cultural Geography*, Conference Proceedings, Department of Geography, University of Edinburgh, 10–12 September.

Rose, Gillian (1993a) 'Some notes towards thinking about spaces of the future', in J. Bird, T. Curtis, T. Putman, G. Robertson, and L. Tickner (eds) *Mapping Futures: Local Culture, Global Change*, New York: Routledge, pp. 70–86.
Rose, Gillian (1993b) *Feminism and Geography: The Limits of Geographical Knowledge*, Cambridge: Polity Press.
Rose, Gillian (1997) 'Spatialities of "community", power and change: The imagined geographies of community arts projects', *Culural Studies*, 11: 1–16.
Rossel, P. (ed.) (1988) *Manufacturing the Exotic*, Copenhagen: International Working Group for Indigenous Affairs.
Roszak, T. (1986) *The Cult of Information*, Cambridge: Lutterworth.
Rushbrook, Dereka (2002) 'Cities, queer space and the cosmopolitan tourist', *GLQ: A Journal of Lesbian and Gay Studies*, 8, 1–2: 183–206.
Ryan, Chris and Hall, C. Michael (2001) *Sex Tourism: Marginal People and Liminalities*, London and New York: Routledge.
Sanders, D. (1995) 'Petition seeks to retain Queen St. parade route', *express: new zealand's newspaper of gay expression*, 14 September: 3.
Sayer, A. (1989) 'Dualistic thinking and rhetoric in geography', *Area*, 21: 301–305.
Sedgwick, Eve (1990) *Epistemology of the Closet*, Berkeley: University of California Press.
Seebohm, Kim (1993) 'The nature and meaning of the Sydney Mardi Gras in a landscape of inscribed social relations', in R. Aldrich (ed.) *Gay Perspectives II: More Essays in Australian Gay Culture*, Sydney: University of Sydney, pp. 193–222.
Seidman, Steven (1996) 'Introduction', in S. Seidman (ed.) *Queer Theory/Sociology*, Cambridge, MA: Blackwell, pp. 1–29.
Selwyn, Tom (1990) 'Tourist brochures as post-modern myths', *Problems of Tourism*, 13: 13–26.
Selwyn, Tom (ed.) (1996) *The Tourist Image: Myths and Myth Making in Tourism*, Chicester: John Wiley.
Shields, Rob (1988) 'Social spatialisation: The case of the West Edmonton Mall', *Environment and Planning D: Society and Space*, 7: 147–164.
Shields, Rob (1991) *Places on the Margin: Alternative Geographies of Modernity*, London: Routledge.
Sibley, David (1995) *Geographies of Exclusion: Society and Difference in the West*, London: Routledge.
Smith, David (1994) *Geography and Social Justice*, Oxford: Blackwell.
Smith, Leighton (1994) *1ZB Leighton Smith*, 8.45am transcript, NewsMonitor Services Limited, Auckland.
Soja, Ed (1992) 'Inside Exopolis: Scenes from Orange County', in M. Sorken (ed.) *Variations on a Theme Park: The New American City and the End of Public Space*, New York: Noonday Press, pp. 94–122.
Sontag, Susan (1966) *Notes on Camp: Against Interpretation*, New York: Delta.
Sontag, Susan (1983) *A Susan Sontag Reader*, London: Penguin.
Spooner, Rachel (1996) 'Contested representations: Black women and the St Paul's carnival', *Gender, Place and Culture: A Journal of Feminist Geography*, 3, 2: 187–203.
Stevens, M. (1995) 'Back to the closet?' *express: new zealand's newspaper of gay expression*, 12 October: 2.
Stoller, Robert (1968) *Sex and Gender*, London: Hogarth Press.
Sunday Star Times (1994) 6 March: 4.

Sunday Star Times (1994) 10 April: A5.
Swain, Margaret (1995) 'Gender in tourism', *Annals of Tourism Research* 22, 2: 247–266.
Swain, Margaret (2004) '(Dis)embodied experiences and power dynamics in tourism research', in Jenny Phillimore and Lisa Goodson (eds) *Qualitative Research in Tourism: Ontologies, Epistemologies and Methodologies*, London and New York: Routledge, pp. 102–118.
Sydney Mardi Gras Parade 1995, ABC Video, Australian Broadcasting Corporation.
Sydney Morning Herald (1998) 'City Search Guide to the Mardi Gras', (http://sydney.citysearch.com.au/E/F/SYDNE/0000/00/64/) (3 March 1998).
Sydney Star Observer (1996) 'Guide to the Mardi Gras festival, parade and party', Number two, 16 February: 12.
Television 3 Network Services Limited (1997) *The 1997 HERO Parade* (programme transcript), Television 3: Wellington.
Torres-Kitamura, M. (1997) 'Out and about', *Hawaii Business*, 42, 10: 23–45.
Urry, John (1988) 'Cultural change and contemporary holiday making', *Theory, Culture and Society*, 5: 509–526.
Urry, John (1990a) *The Tourist Gaze: Leisure and Travel in Contemporary Societies*, London: Sage.
Urry, John (1990b) 'The consumption of "tourism"', *Sociology*, 24: 23–35.
Urry, John (1992) 'The tourist gaze "revisited"', *American Behavioural Scientist*, 36: 172–186.
Vaiou, D. (1992) 'Gender divisions in urban space: Beyond the rigidity of dualist classifications', *Antipode*, 24: 247–262.
Valentine, Gill (1996) '(Re)negotiating the "Heterosexual street": Lesbian productions of space', in Nancy Duncan (ed.) *BodySpace: Destabilizing Geographies of Gender and Sexuality*, London and New York: Routledge, pp. 146–155.
Valentine, Gill and Skelton, Tracey (2003) 'Finding oneself, losing oneself: The lesbian and gay scene as paradoxical space', *International Journal of Urban and Regional Research*, 27, 4: 849–866.
Van der Meer, Theo (1997) 'Sodom's seed in the Netherlands: The emergence of homosexuality in the Early Modern Period', *Journal of Homosexuality*, 34: 1–16.
Veijola, Soile and Jokinen, Eeva (1994) 'The body in tourism', *Theory, Culture and Society*, 11: 125–151.
Waitt, Gordon (2003) 'Gay games: performing "community" out from the closet to the locker room', *Social and Cultural Geography*, 4, 2: 167–183.
Walker, Lisa (1995) 'More than just skin deep: Fem(me)inity and the subversion of identity', *Gender, Place and Culture: A Journal of Feminist Geography*, 2: 71–76.
Walker, M. (1991) 'Disneyfication of the planet', *The Guardian Weekly*, 4 August: 21.
Wichtel, D. (1997) 'Television review: Better later than never', *New Zealand Listener*, 22 March: 64.
Wilson, Elizabeth (1988) *Hallucinations: Life in the Post–Modern City*, London: Radius.
Wolff, Janet (1995) *Feminist Sentences*, Oxford: Blackwell.
Yanagisako, S. and Collier, J. (1990) 'The mode of reproduction in anthropology', in D. Rhode (ed.) *Theoretical Approaches on Sexual Difference*, New Haven and London: Yale University Press, pp. 131–141.
Young, A. (1990) *Femininity in Dissent*, London: Routledge.

Young, Iris Marion (1990) *Justice and the Politics of Difference*, Princeton, NJ: Princeton University Press.
Zukin, Sharon (1991) *Landscapes of Power: From Detroit to Disney World*, Berkeley: University of California Press.

Websites

http://gaytravelnet.com/aus/Sydney.html (accessed 14 May 2004)
http://www.edinburghfestivals.com (assessed 30 May 2004)
http://www.marchingnz.org.nz
http://welcomehome.org/rainbow/lists.html
http://electriciti.com/tvlexprs/pages/menvac.htm

Index

Italic page numbers indicate figures and illustrations not included in the text page range.

AAG (Association of American Geographers) 32
abjection 7, 11, 26–8, 54, 64–6, 75, 124–5
Above and Beyond tours: travel agents 109
ACT UP 105
Adam, Barry 4
Adelaide Grand Prix 16
advertising 103, 106, 109–10
AIDS *see* HIV/AIDS
Aldrich, Robert 117
all-women drumming group 6, 33–42, 115–16, 124
Annals of Tourism Research 19
Antipode 32
Aotearoa/New Zealand: acceptable parades 84–5; census 69; human rights legislation 78; newspapers 87; Prime Minister 62; *see also* HERO parades
Armour, Lawrence 113
assimilation 25, 68, 114
Association of American Geographers (AAG) 32
Auckland: camp performances 124; HERO parade debate 78–99; 'homonormativity' 125; parade opposition 8; queer spaces 101; and Sydney 106; tourists' responses 69–70; *see also* HERO parades
Auckland Coming Out Day Parade 4–5
Australia 69, 110; *see also* Sydney Gay and Lesbian Mardi Gras Parade

barriers 58–62 *see also* borders; boundaries
Batchelor, Julian 86

Baudrillard, Jean 17
Becker, Toby 117, 119
Belk, Russell 113
Bell, David 13, 20, 29, 30, 31, 36, 39, 45, 51, 56, 57, 64, 101, 104, 105, 114
Bennie, J. 44
Berg, Larry 14, 77
Berlusconi, Silvio 118
Binnie, Jon 12, 17, 20, 28, 29, 31, 100, 104, 106, 110, 113–14, 124, 126
biopower 45
Blum, Virginia 32–3
bodies: definition of 11; displays 6–7; gendering and sexualising of 2–3, 31–2; materiality of 22–5; queer 10–15, 57; research on 3; spectacle of 9, 57; 'ugly' 7, 27, 54, 65–6, 75, 125
Bodies: Exploring Fluid Boundaries 2
body building 43
body fluids 12
BodySpace:Destabilizing Geographies of Gender and Sexualities 21
bondage performance 54
Bondi, Liz 36, 115
Bonnemaison, S. 56
borders: corporeal 54–5, 63; instability of 54, 124–5; physical 58–62, 75
Bordo, Susan 14, 30
boundaries 5, 8; *see also* barriers; borders
Brickell, Chris 58, 77
Brighton 17
Brighton Ourstory Project 17
Brimble, Dennis 90
Britton, Steven 103
Brown, Michael 17, 101
Bruto, L. G. 118

Butler, Judith 22, 28, 29, 36, 41, 52, 124, 126–7

Cady, Joseph 118
Caldwell, F. 89, 92, 93, 94
Callard, Felicity 12
Cameron, E. 88
camp 11, 28, 31, 34, 35–6, 43, 45, 124–5
Carberry, G. 111, 112
carnivals 23–4, 55–6
Carter, Simon 20
Castaneda, Quetzil 23
Catholicism 102, 117, 121
CBD (commercial and business district): Auckland 78–99
celebratory tone 100–1
Chouinard, Vera 21
Christie, Nigel 89
Chung, Robert 44
Cieri, Marie 100
citizenship 1, 8, 114, 125
city tourism 8
Cleto, Fabio 28
Clift, Stephen 20
Cloke, Paul 17
club scene 104
Cohen, A. 56
Cohen, E. 17, 18, 20
Cohen, Roger 104
Collier, J. 18
commercialisation 100, 111, 113, 114, 126
conferences: 'Gender/Tourism/Fun?' 24
conservatism 114, 118
consumer capitalism 33
corporate links: USA 113
costumes 4, 6, 34, 42, 51, 105, 119
Craik, Jennifer 16
Cream, Julia 11, 12, 15, 29, 52
Cresswell, Tim 56
Crowley, Jo 92–3
cultural imperialism 63, 73, 75

Dahir, Mubarak 113
data collection methods 6
De Lauretis, Teresa 3, 25–6, 123
Del Casino, Vincent 103
Demon float 83–4
Derrida, Jacques 13
Descartes' theory of separation 14
differences 100–1
Disneyland 18
Doc Marten boots 51
Douglas, Mary 62

drag queens 105, 119
drumming groups 6, 33–42, 116, 124
dualistic thinking 10, 13–15, 77, 98, 122, 125
Duncan, Nancy 21, 56
Dunedin Festival parade 4, 5
Dyer, Richard 28, 45

Eagleton, T. 56
Eco, U. 18, 112
Edensor, Tim 37, 42, 96, 103, 110
Edinburgh 6, 36–7, 101–2, 114–17, 121, 124, 126
Edinburgh drumming group 6, 33–42, 115–16, 124
Edinburgh focus groups: first 34, 37–8, 40, 116; second 39
Ely, L. 96, 97
embodiment 6–7, 11–15, 22–5
the Enlightenment 14
Enloe, Cynthia 19
equal rights 66, 82, 114, 117, 121, 125
ethnicity 4, 19, 20, 66, 69
express Check-Out Chicks 48–9

Featherstone, M. 18
Feifer, M. 17
femaleness 13–14
feminised images 19
feminist geographers 3
feminists 18, 21
feminist theorists 2
Fisher, A. 16
flânuer (lesbian) 35–6
floats: 1994 HERO Parade 80; 1996 HERO Parade 63–6, 70–4, 83–4; Sydney Mardi Gras Parade 59
Foucault, Michel 13, 21–2, 45, 58
Fox, T. C. 118

GABA (Gay Auckland Business Association) 7–8, 74–5
Gaily Normal 7–8, 72–4
Game, Ann 22
Gay and Lesbian Quarterly (GLQ) 24
gay and lesbian tourism 2, 20–5, 100, 110
Gay Auckland Business Association (GABA) 7–8, 74–5
gay liberation movement 4
gay male sex 64–5
gay neighbourhoods 17, 104, 105, 112–13, 118; *see also* Ponsonby, Auckland

gay pride parades: abjection 63–7; affects of 100–1; carnivals 55–6; controversy 77; *Gay and Lesbian Quarterly* papers 25; general 1; identity and pride 3–6; performativity 29; politics of 105–6; promotion 108–11; purpose 23; and queer bodies 2; queer enclaves 61–2; road barriers 58–61; safety issues 58–61; as spectacles 104; urban space 56–8; *see also* drumming groups; HERO parades; Sydney Gay and Lesbian Mardi Gras Parade; World Pride Roma
gay rights 107, 118
gay skinheads 30, 31
gay tourism *see* gay and lesbian tourism
gay tourists 61–2, 110; questionnaire responses 69–70
gay travel marketing 10, 25, 108–11
gay villages 104
gaze: of tourists 9, 15, 23, 106
Gearing, Nigel 72, 73, 80, 85
gender 3, 12–13, 18–19, 23, 29
gender/sex distinction 18–19, 21–2
Giorgi, Gabriel 24
Giubileo (Vatican's Jubilee) 102, 117, 121, 126
Glasgow 115–16
globalization 24, 25
GLQ (Gay and Lesbian Quarterly) 24
'God is gay' poster *120*
Grant, Ali 21
Greenwich Village 4
Greenwood, D. 10
Gregory, Derek 18
Gregson, Nicky 29
Grosz, Elizabeth 2, 10, 11, 14, 15, 19, 31–2, 43, 44, 62, 64
group oppression 63, 66

Hall, C. Michael 16, 20, 56
Hall, Derek 19
Hanna, Stephen 103
Haraway, Donna 16
Harvey, David 18
Hatch, J. 16
Hay, David 80, 82, 83–4, 85, 86–7
heritage attractions, Stirling 103
heroes, sporting 80–1
HERO Festival 5–6, 56, 86, 93
HERO Marching Boys 42–52
HERO Marching Girls 51–2
HERO parades: abjection 64–6; borders 58; corporeal borders 63–4; floats 70–5; performance 106–8; promotion 108–11; road barriers 60–1; site of 78–99, 125; subversion 121; tourists' responses 63, 67–8
HERO Project Director 57, 91, 93–4, 95, 97, 106–8
heterosexuality 21–2, 25–6, 68–9; parody of 6, 8
heterosexual tourists 106, 110
Hill, Kim 9, 82, 87
Hindu festival *(thaipusam)* 22–3
HIV/AIDS 17, 20, 64–5, 106–7
Hodge, Stephen 98
Holcomb, B. 20
Holmes, Paul 83
'homonormativity' 1, 101, 121, 125
homosexuality 21–2, 25–6, 85–6, 113–14, 117–18, 125
homosexual oppression 1, 28, 50, 63, 115
'hoon effect' 16–17
Hubbard, Phil 101
Huberman, A. M. 6
Hughes, Howard 20
human rights legislation 78
humour 35, 40, 53
hyperfeminine lipstick lesbian 30, 31, 39
hypermasculine skinhead 30, 31
'hyperreal' spaces 18

identity–pride connection 4
Ieriko, Paulo 96
inequality 18, 66
international queer tourism sites 101
international tourists 110
Italian Gay Pride March (1994) 118; *see also* World Pride Roma

Jackson, Peter 23, 56
Jagose, Annamarie 3, 11, 124
James, Bev 81
Jayne, Mark 101
Jenkins, Andrew 20
Johnson, Louise C. 15, 21
Jokinen, Eeva 3, 10, 15, 16, 20, 22, 23, 36, 123, 126

Kantsa, Venetia 24
Kates, Steve 113
Keith, H. 86
Kinnaird, Vivian 19
Kippendorf, J. 22

Kirby, Vicky 22–3, 30
Knopp, Larry 30, 36, 111, 114, 115
Kothari, Uma 96, 103, 110
Kristeva, Julia 7, 26, 27, 54, 62, 64, 65, 75

La Rocca, O. 119
Legat, Nicola 81–2, 83, 85–6, 87
lesbian bodies 64
lesbian feminists 51
lesbians: anonymity 40–1; 'butch' 34; and domestic spaces 3; flânuer 35–6; lipstick 30, 31, 39, 105; tourists' reaction to 64
lesbian tourism 20–1, 24–5, 33
Lesvos 24
Levine, M. 18
Lewis, Claire 23–4, 55, 56
Ley, D. 112
liberal humanism 66–7
liberalism 7, 76
lipstick lesbians 30, 31, 39, 105
Lloyd, Genevieve 13, 14
Locker Room Boys 42–3, 48
London 56
London Pride 107
Longhurst, Robyn 2, 11, 12, 13, 14, 15, 77, 101
Luongo, Michael 20, 25, 117, 118

Madrid 24
Mahtani, Minelle 126–7
male marching groups 42–51
maleness 13–14
Manning, F. E. 56
man/woman binary 13–15
marching groups 6, 7, 32, 124; see also all-women drumming group; men's marching groups
marginal places 17
Markwell, Kevin 25, 109–10
marshals 60–1
Marshment, M. 19
masculinist knowledge 14–15, 123–4
masculinist rationality 14, 16
Massey, Doreen 99, 101
MacCannell, Dean 10, 17, 22, 24
McClintock, Anne 27
McDowell, Linda 77
McNaught, Anita 46, 50–1
McNeill, Donald 118
media coverage 36–7
Men on Vacation: travel agents 108–9
men's marching groups 42–51

Miles, M. 6
Mills, Lee 87
mind/body dualism 14–15, 19, 77, 78
Montreal Parade 51–2, 104–5
Moore, L. 6, 97
Morgan, Nigel 20
Morin, Karin 13
Morris, Meghan 19
Mulvey, Laura 19
Munt, Ian 18
Munt, Sally 35, 36, 39, 41, 56
Myslik, Wayne 21

Namaste, Kim 20, 101, 104, 105
Nast, Heidi 12, 32
Nelson, Lise 101
neocolonialism 100
neoliberal cultural citizenship 1, 125
newspapers 36–7
New York Stonewall riots 4, 107, 112
New Zealand see Aotearoa/New Zealand
Nietzsche, Frederick 13
Notting Hill Carnival 55–6

Olds, K. 16, 112
Oliver, Kelly 27
onlookers 37–9, 49–50, 57, 61–2
oppression 1, 28, 39, 50, 51, 63, 66, 115; see also self-oppression
the Other 10, 15–17, 20–1, 23, 61, 62, 63, 122–3, 126–7

Paix, P. 16
Pākehā 101
parade entries 7–8
parade officials 59–61
parades 3–6; acceptable 84–5; see also gay pride parades
parade sponsors 96
paradoxical performances 126–7
paradoxical space 123
patriarchy 32, 51
Peace, Robin 13
performance 23, 29–30, 107; see also HERO parades; marching groups; Sydney Gay and Lesbian Mardi Gras Parade
performances 6, 42, 54, 124–5, 126
performativity 11, 29–30, 31
Perkins, Harvey 17
Perth Parade: Australia 116–17
Phelan, Shane 114
Phillips, Jock 81
philosophy: early Greek 13–14

Piggford, George 49
Pile, Steve 12, 23–4, 55, 56
'pink pound/dollar' 20
Plato 13–14
Podmore, Julie 62
Poggioli, Sylvia 119
Ponsonby, Auckland 8, 78–9, 88–99, 101, 121, 125
postmodernism: and tourism 17–18
power relations 34
pride parades *see* gay pride parades
Pride Scotland 6, 34, 42, 101–2, 115, 116, 121, 124, 126
Prime Minister: of Aotearoa/New Zealand 62
Princes Street, Edinburgh 37–9
Pritchard, Annette 20, 21
Probyn, Elspeth 30, 39, 50, 51–2
promotion 108–10
protest 4, 34, 55, 66, 78, 84–6, 98, 107, 112, 125
Puar, Jasbir Kaur 21, 24, 25, 33, 100, 124
Pythagoreans 13

Queen Street 78–99, 125
queer activists 104–5
queer bodies *see* bodies
Queer Nation 105
queer theory 3, 11, 25–30, 126; *see also* abjection; camp; performativity
queer tourism studies 22–5, 33, 100
queer tourists 61–2, 70
questionnaires 6, 7, 54, 57, 61, 63, 67–70

race 23, 25, 32, 56, 66, 88, 100–1, 123, 126
racism 27, 32, 66
Rainbow Youth organisation 7–8, 70–2, 90–2, 125
rationality: masculinist 14, 16
religious rituals 22–3
research *see* tourism studies
Richardson, Diane 100, 106, 114, 125
Rio Carnival 24
Rise of the Gay and Lesbian Movement, The 4
road barriers 58–61
Roche, M. 16
Rogers, Warwick 75
Rojek, Chris 22, 77
Rome 101, 102, 117–19, *120,* 121, 126
Rose, Gillian 15, 16, 29, 31, 50, 77, 123
Rossel, P. 10

Roszak, T. 17
routines, marching/dance 33, 42; *see also* performances
rugby players 44, 80, 81
Rushbrook, Dereka 24–5, 104
Rutelli, Francesco 118
Ryan, Chris 20

sadomasochism performance 7, 54, 61, 63, 125
safe sex: billboards 94, *95,* 96–7; floats 63, 65, 80, 83, 85
safety issues 58–61
Sanders, D. 88
Saville-Smith, Kay 81
Sayer, A. 77
Scotland 103, 115; *see also* Pride Scotland
Sedgely, Diane 20
Sedgwick, Eve 101
Seebohm, Kim 112–13
Seidman, Steven 26
self-mockery 28
self-oppression 73
self/Other 10, 54–5, 62, 72, 122–3
Selwyn, Tom 17
separation: theory of 14
sex: as a term 18–19
Sex and Gender 18
sex tourism 20–1; *see also* gay and lesbian tourism
sexual identities 29–30, 31, 53
sexuality: of bodies 11–13; and gender 31–3, 36, 53, 114; and queer theory 3; and tourism 20–1, 104
Sexuality and Space Specialty Group (SSSG) 32
shame 40–1, 127
Shields, Rob 17, 18
Shipley, Jenny 62
Sibley, David 27
Skelton, Tracey 123
Smith, David 66
Smith, Leighton 82–3
social control 58, 112–13
social effects: of events 16
Sodano, Angelo 119
Soja, Ed 18
Sontag, Susan 35, 36
souvenirs 108
spacial segregation 58, 61–2
Spain 24
spectacles 57, 58, 106; of bodies 9, 44, 50, 99

Index 145

spectators 37–9, 49–50, 57, 61–2
Spooner, Rachel 23, 56
SSSG (Sexuality and Space Specialty Group) 32
Stevens, M. 88, 89
Stirling, Scotland 103
Stoller, Robert 18
Stonewall riots 4, 107, 112
street festivals 23–4
streets: queering of 56–8
subjectivity 52, 123
Swain, Margaret 19–20
Sydney 9, 101, 111, 121, 124, 126
Sydney Gay and Lesbian Mardi Gras Parade: affirmation of Other 112; barriers 58–60; festival events 111; *Gay and Lesbian Quarterly* papers 25; general 5–6; 'homonormativity' 1, 121; map 102; marching groups 42–3; performance 107; political impact 56; promotion 108–11; queer tourists 62; site 101; souvenirs 58; spectators 57; tourism 99; tourists' responses 9, 67–8

television 108
thaipusam (Hindu festival) 22–3
theory of gender 29
theory of separation 14
tolerance 68
Topp, Lynda and Jools 50–1
Torres-Kitamura, M. 20
tourism research 7, 13, 16–25
tourism studies 1–2, 10–11, 15–25, 33, 103–4, 122–3
tourists: gay 61–2, 69–70; gaze of 23; questionnaires 6, 7, 54, 57, 61, 63, 67–70; responses of 7, 54, 67–8
tourist sites 77, 101, 103
tour operators *see* travel companies
training 47–8
transgenders 65–6

Transpride float 65–6
travel brochures 19
travel companies 9, 108–11
travel industry 21
travel marketing 10

'ugly bodies' 7, 27, 54, 65–6, 75, 125
'uniforms' 42
United Kingdom (UK) 113–14
Urry, John 15, 17, 18, 22
USA: Stonewall riots 4, 107, 112

Vaiou, D. 77
Valentine, Gill 3, 20, 21, 29–30, 56, 57, 64, 101, 104, 105, 123, 126
Van der Meer, Theo 117
Vatican 117, 118, 119
Vatican's Jubilee *(Giubileo)* 102, 117, 121, 126
Veijola, Soile 3, 10, 15, 16, 20, 22, 23, 36, 123, 126
VIP area: 1998 HERO Parade 62

Waitt, Gordon 111
Walker, Lisa 30
Walker, M. 18
Wolff, Janet 19
women: construction of 36, 115; as different 18–20; oppression of 51; and philosophy 13–15; and queer tourism 24
women's drumming group 6, 33–42, 115–16, 124
World Pride Roma 6, 25, 102, 117–19, 120, 121, 126

Yanagisako, S. 18
Young, A. 51
Young, Iris 7, 27, 54, 62, 63, 65, 66, 67, 68, 74, 75, 125

Zukin, Sharon 18